20th Annual Conference and 9th International Conference of Chinese Society of Micro/Nano Technology (CSMNT2018)

Journal of Physics: Conference Series Volume 1209

Medan, Indonesia
12-14 December 2018

ISBN: 978-1-5108-9548-5
ISSN: 1742-6588

Printed from e-media with permission by:

Curran Associates, Inc.
57 Morehouse Lane
Red Hook, NY 12571

Some format issues inherent in the e-media version may also appear in this print version.

Institute of Physics
Dirac House, Temple Back
Bristol BS1 6BE UK

Phone: 44 1 17 929 7481
Fax: 44 1 17 920 0979

techtracking@iop.org

Additional copies of this publication are available from:

Curran Associates, Inc.
57 Morehouse Lane
Red Hook, NY 12571 USA
Phone: 845-758-0400
Fax: 845-758-2633
Email: curran@proceedings.com
Web: www.proceedings.com

TABLE OF CONTENTS

IOP Conf. Series: Journal of Physics: Conf. Series **1209** (2019) 011001 doi:10.1088/1742-6596/1209/1/011001

Preface

This book collects papers selected from the 20th Annual Conference of the Chinese Society of Micro/Nano Technology and CSMNT 9th International Conference of Micro/Nano Technology (CSMNT2018).

The rapid development of micro-nano technology speeds up the industrialization of micro-and nona-research results, and the industry has achieved remarkable benefits. With the deep inregration of technology and science research, the knowledge of micro-nano has been learnt and understood by more and more people. Micro-nano technology has brought tremendous changes in many fields including communications, transportation, healthcare, energy, environment, etc.

CSMNT2019 hold in Zhengzhou with topic "Micro-Nano technology makes our life better". The conference invites scientists in micro-nano technology all over the world to discuss the state-of –the-art and the development of the micro-nano technology. The conference set up a variety of activities such as general report, session report, poster exchanging, technical exhibition and technological innovation contest. The conference focused on the frontier of science and technology both in China and abroad, and target to improve people's live-hood by encourage the industrialization of the technology.

There are many top level MEMS and NANO technology researchers were invited to make plenary presentations. For example, NAE member, **Stephen Z. D. Cheng** gave a speech named ***Giant Molecules based on Nano-atoms***; The National Academy of Engineering of Korea, **Chul B. Park** gave a speech named ***Nano-Fibril Technology for Superior Properties and Foaming-Ability***, and 12 more excellent specialist gave the brilliant talk.

Thanks the staff of Zhengzhou University and CSMNT Secretary for the efforts they have made, and thanks all honored participants. Their contributions in all aspects are indispensable to the conference's success.

Thanks to IOP Publishing for their work on publishing the special volume for the selected papers of this conference. We have good experience for cooperation, and hope that we can continue this cooperation in coming conferences in the future.

IOP Conf. Series: Journal of Physics: Conf. Series **1209** (2019) 011001 doi:10.1088/1742-6596/1209/1/011001

The 21st Annual Conference of the Chinese Society of Micro/Nano Technology and CSMNT 10th International Conference of Micro/Nano Technology (CSMNT2019) will be held on 11-14th, Oct., 2019, Wuhan, China.

Welcome to Wu Han!

Zheng You
Chairman of Organizing Committee
President of Chinese Society of Micro/Nano Technology

IOP Conf. Series: Journal of Physics: Conf. Series **1209** (2019) 011002 doi:10.1088/1742-6596/1209/1/011002

Peer review statement

All papers published in this volume of *Journal of Physics: Conference Series* have been peer reviewed through processes administered by the proceedings Editors. Reviews were conducted by expert referees to the professional and scientific standards expected of a proceedings journal published by IOP Publishing.

IOP Conf. Series: Journal of Physics: Conf. Series **1209** (2019) 012001 doi:10.1088/1742-6596/1209/1/012001

Design of piezoelectric self-sensing microinjection device

X W Wei [1,2], L P Sun [2,*], C Su [1], Y L Shi [1] and J Q Li [2]

[1] Electrical and Information Engineering Department,Heilongjiang University Of Technology, china
[2] College of Mechanical and Electrical Engineering,Northeast Forestry University, china

wxw198806@163.com, zdhslp@163.com, jxdxsc@163.com, duchunxiaoeasy22@163.com, lijiaqi9306@163.com

Abstract. With the development of biotechnology and injection technology, cell micro flow syringes have been widely used in many fields. In the meanwhile, the requirements of experimental equipment and experimental method are becoming higher and higher. In this paper, the author study on the theory and experiment of the first direct and secondary converse piezoelectric effects of piezoelectric ceramics by using the piezoelectric characteristics which can realize the integration of sensors and actuators. And the author design a self-sensing micro injection device which has the functions of localization, injection output and self-perception. The control precision and operation have been effectively improved by means of analyzing the relationship between the output and the input.

1. Introduction

With the continuous extension and development of nanotechnology, people begin to conduct a in-depth study of macroscopic and microscopic things. They not only explore some phenomena on the surface of things, but also make progress in further the development of miniaturization and precision. The scientific, medical and even government circles have put forward more and more high requirements to the MEMS system. A variety of piezoelectric sensors and actuators can be made by the single piezoelectric effect of piezoelectric materials. However, this separation sensor and actuator have been long existed and can not meet the needs of high technology. Therefore, a new type of micro sensor and micro actuator is extremely urgent by using the multiple piezoelectric effect theory[1,2].

Because micro processing technology itself has many advantages, more and more industry begun to take advantage of the technology to develop all sorts of detection device, such as fluid monitoring, super precision machining, biological genetic engineering and so on. Microinjection device is one of them. In the piezoelectric applications, most of microinjection device use first converse piezoelectric effect of piezoelectric materials to complete the executive function, and does not apply multiple piezoelectric effects. The purpose of this study is using the first converse piezoelectric effect of actuator and the secondary direct piezoelectric effect of sensor in the same piezoelectric body to design an automatic microinjection device which can sense the actual liquid flow. While applying an external voltage on piezoelectric materials, it will output a small displacement. Use the secondary positive piezoelectric effect of our own to perform sensor functions. Use part of the output charge to sense the actual amount of microinjection produced by the inverse. In the case of the amount of injection, the value of real injection can be realized. The control precision and operation efficiency of the microinjection device are improved effectively and the synchronism between the sensor and the actuator is reflected. Most microinjection devices will appear wear and deformation for a long period of time resulting in the output error between the theoretical value and the actual value. Self-sensing microscopic flow injection device can more accurately measure the actual output displacement using multiple piezoelectric effect, so that they can better control the dosage of micro syringe. The injection rate improve greatly. Therefore, it is great significance to study the principle and structure of the self-sensing microinjection device for the optimization of the original injection device[3-5].

2. The basic principle of piezoelectric self-sensing technology

2.1. Theoretical basis of piezoelectric self-sensing technology

Piezoelectric self-sensing technology refers to the use of piezoelectric ceramic materials of the direct piezoelectric effect and converse piezoelectric effect at the same time and piezoelectric ceramic output displacement or force for actuator to achieve self-sensing at the same time. The direct piezoelectric effect is that when the piezoelectric ceramic material is subjected to external forces in a certain direction, its surface will generate free charge, as shown in figure 1(a). The converse piezoelectric effect is that the surface of piezoelectric ceramic generates a free charge when an external force is applied in a particular direction, as shown in figure 1(b). For piezoelectric ceramic materials, the positive piezoelectric effect and inverse piezoelectric effect are generally simultaneous and reversible, as shown in figure 2. The primary inverse piezoelectric effect on the same piezoelectric ceramic material is used to realize the sensor function and the secondary positive piezoelectric effect is used to realize the actuator function. It can be regarded as the realization process of piezoelectric self-perception[6].

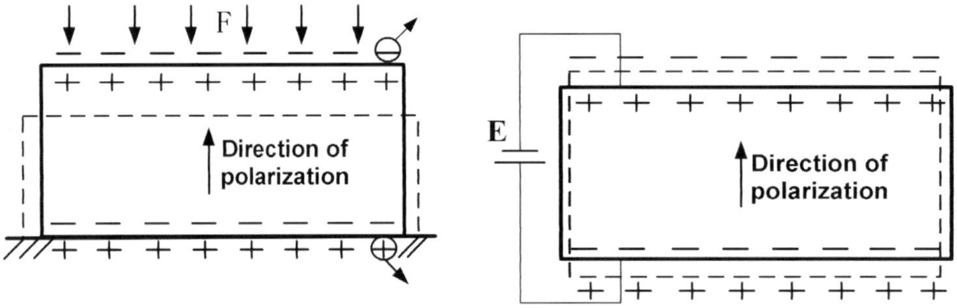

(a) direct piezoelectric effect **(b)** converse piezoelectric effect

Figure 1. Piezoelectric effect.

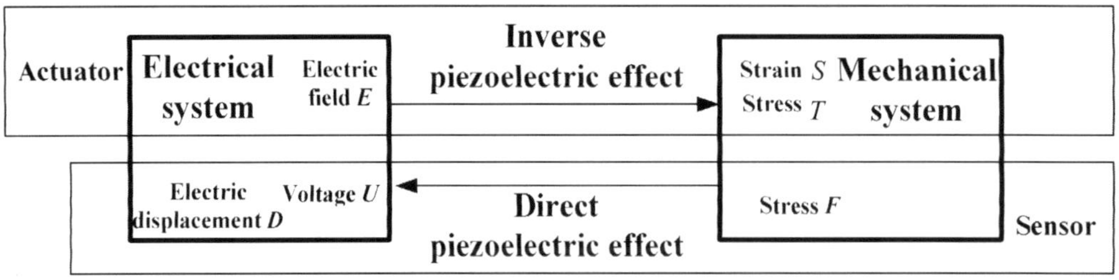

Figure 2. Principle of piezoelectric self-sensing.

The electrical and mechanical states of piezoelectric materials are determined according to the form of boundary conditions. Therefore, the boundary conditions of piezoelectric materials play an important role in the study of piezoelectric effects. From the angle of "machine-electrical" coupling, piezoelectric body can be under the condition of electrical and mechanical boundary conditions. Electrical boundary conditions are divided into electrical open circuit and short circuit. And mechanical boundary conditions are divided into mechanical freedom and clamping. Four types of combined boundary conditions can be obtained by combining the electrical and mechanical boundary conditions, and four types of piezoelectric equations can be obtained under the combined boundary conditions. The specific expression method is shown in table 1[7].

Table 1. Expressions and boundary conditions of four kinds of piezoelectric equation.

Piezoelectric equations	Expression	Boundary conditions	Remarks
The first (d-type)	$S_i = s_{ij}^E T_j + d_{ni} E_n$ $D_m = d_{mj} T_j + \varepsilon_{mn}^T E_n$	Mechanical freedom Electrical short circuit	s_{ij}^E :Short circuit elastic compliance constant matrix d_{ni} :Transposed matrix of d d_{mj} :Piezoelectric strain constant matrix ε_{mn}^T :Dielectric constant matrix
The second (e-type)	$T_j = c_{ji}^E S_i + e_{nj} E_n$ $D_m = e_{mi} S_i - \varepsilon_{mn}^S E_n$	Mechanical clamping Electrical short circuit	c_{ji}^E :Short circuit elastic stiffness constant matrix e_{nj} :Transposed matrix of e e_{mi} :Piezoelectric strain constant matrix ε_{mn}^S :Clamping dielectric constant matrix
The third (g-type)	$S_i = s_{ij}^D T_j + g_{mi} D_m$ $E_n = -g_{nj} T_j + \beta_{nm}^T D_n$	Mechanical freedom Electrical open circuit	s_{ij}^D :Open circuit elastic stiffness constant matrix g_{mi} :Transposed matrix of g g_{nj} :Piezoelectric voltage constant matrix β_{nm}^T :Free dielectric isolation matrix
The fourth (h-type)	$T_j = c_{ji}^D S_i - h_{mj} D_m$ $E_n = -h_{ni} S_i + \beta_{nm}^S D_m$	Mechanical clamping Electrical open circuit	c_{ji}^D :Open circuit elastic stiffness constant matrix h_{mj} :Transposed matrix of h h_{ni} :Piezoelectric stiffness constant matrix β_{nm}^S :Clamp dielectric isolation rate matrix

Under mechanical freedom and only applied external electric field E_n^w, the expression can be obtained according to the first converse piezoelectric effect and the first piezoelectric equation.

$$S_i^{(1)} = d_{ni} E_n^w \tag{1}$$

The simplified Eqs(2) is obtained by Eqs(1)[8].

$$\delta^{(1)} = nd_{33}V \tag{2}$$

Where, δ is the first converse output displacement, n is the layer number of piezoelectric ceramic stack, d_{33} is the piezoelectric coefficient, V is the driving voltage.

The strain $S_i^{(1)}$ generated in the first converse process will react on piezoelectric ceramics, the expression can be obtained according to the secondary direct piezoelectric effect and the second piezoelectric equation.

$$D_m^{(2)} = e_{mi}S_i^{(1)} = e_{mi}d_{ni}E_n \tag{3}$$

The simplified Eqs(4) is obtained by Eqs(3)[9].

$$V^{(2)} = \frac{Q^{(2)}}{C} = \frac{AD^{(2)}}{C} = \frac{Ae_{33}d_{33}V_3}{Ct} = \frac{Ae_{33}}{Ct}\delta \tag{4}$$

Where, $V^{(2)}$ is the output voltage of the secondary direct piezoelectric effect, C is the equivalent capacitance of piezoelectric ceramics, A is the section area of piezoelectric ceramics, t is the thickness of a single piece ceramic. And $e_{33} = d_{33}y_{33}$, e_{33} is the piezoelectric stress constant, d_{33} is the piezoelectric coefficient, y_{33} is the elastic coefficient.

It can be seen from the above analysis that the displacement generated by Eqs(2) is used to control the stroke and injection amount of the microinjection device, and the voltage generated by Eqs(4) is used to perceive the actual output displacement.

2.2. Experimental of piezoelectric self-sensing technology

The main equipment used in the experiment are PZT-5, YE6232B data acquisition, YE5850 charge amplifier, XE-500/501 piezoelectric controller, MDS amesdial, load cell and computer. The basic parameters of PZT are shown in table 2.

Table 2. Basic parameters of PZT-5.

Type	Piezoelectric coefficient (10^{-12} C/N)	Piezoelectric voltage constant (m^2/C)	Piezoelectric stress constant (C/m^2)	Elastic coefficient (10^{10} N/m^2)	Piezoelectric stiffness constant (10^9 N/C)
PZT-5	$d_{33} = 800$	$g_{33} = 0.019$	$e_{33} = 34.4$	$c_{33} = 4.3$	$h_{33} = 2.15$
	Number of pieces	Thickness of the single chip (mm)	Capacitance (μF)	Maximum driving force (N)	Boundary dimension (mm)
	250	0.12	1.75	4000	$10\times10\times30.4$
	1000	0.12	1.75	1000	$5\times5\times121.6$

The experimental structure and physical connection of the first converse piezoelectric effect are shown in figure 3 and figure 4.

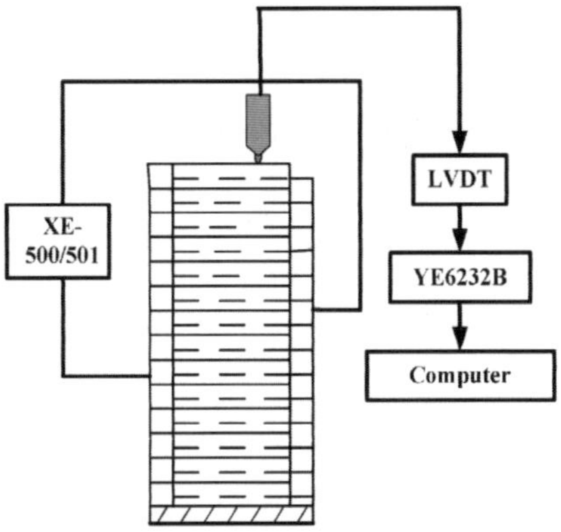

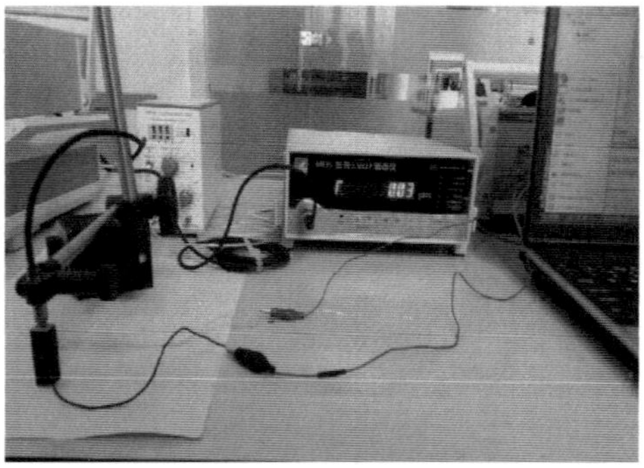

Figure 3. Experimental structure diagram. **Figure 4.** Physical connection diagram.

The piezoelectric ceramics are applied from 0V to 100V at a voltage interval of 10V, the experimental measurement and theoretical output values of the micro displacement generated by the first converse piezoelectric effect are shown in table 3.

Table 3. Practical value and theoretical value.

Driving voltage (V)	10*10*30.4		5*5*121.6	
	Theoretical value (μm)	Practical value (μm)	Theoretical value (μm)	Practical value (μm)
0	0	0.013	0	0.0267
10	2	1.477	8	5.863
20	4	3.453	16	14.017
30	6	5.683	24	20.863
40	8	8.113	32	29.087
50	10	10.333	40	37.040
60	12	12.727	48	45.087
70	14	15.247	56	52.963
80	16	17.513	64	60.890
90	18	18.310	72	72.747
100	20	21.773	80	80.653

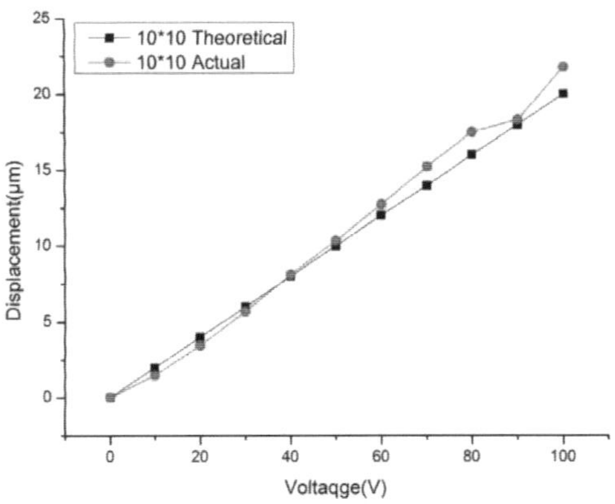

(**a**) 10*10*30.4

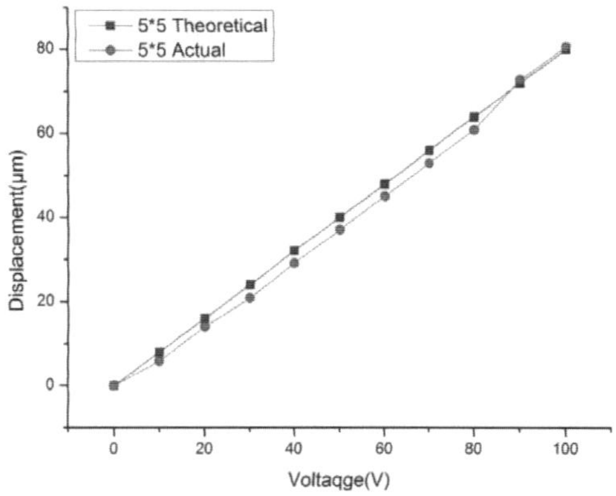

(**b**) 5*5*121.6

Figure 5. Comparison diagram of theoretical and practical.

It can be seen from the figure that the applied driving voltage is basically in a linear relationship with the output micro displacement. The measured values of 10*10*30.4 stack and 5*5*121.6 stack between 0V~80V basically coincide with the theoretical values respectively. And the resulting error can be ignored. In practical application, the driving voltage of the two models should be controlled between 0V~80V to improve the system precision.

The experimental structure and physical connection of the secondary direct piezoelectric effect are shown in figure 6 and figure 7.

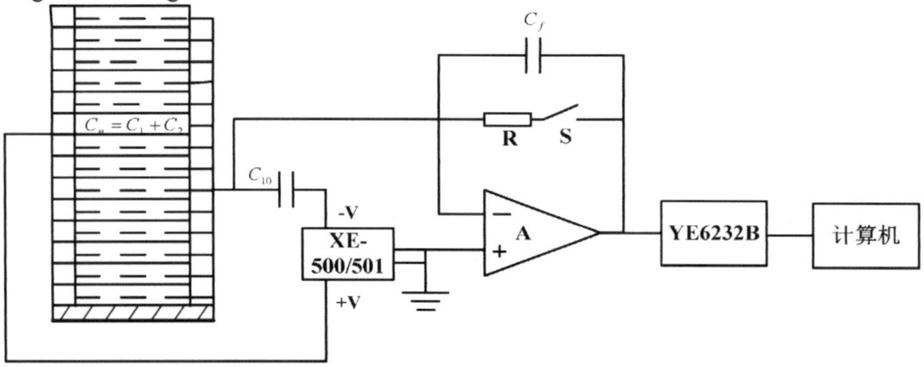

Figure 6. Experimental structure diagram.

Figure 7. Physical connection diagram.

The experimental measurement and theoretical output values are shown in table 4.

Table 4. Practical value and theoretical value.

Driving voltage (V)	10*10*30.4		5*5*121.6	
	Theoretical value(mV)	Practical value(mV)	Theoretical value(mV)	Practical value(mV)
0	0	0	0	0
10	32.76	10.88	32.76	14.13
20	65.52	24.37	65.52	27.67
30	98.28	36.58	98.28	41.27
40	131.04	44.33	131.04	54.69
50	163.8	57.78	163.8	69.45
60	196.56	72.79	196.56	82.76
70	229.32	85.08	229.32	96.42
80	262.08	97.30	262.08	110.27
90	294.84	113.48	294.84	127.89
100	327.6	125.77	327.6	141.39

According to table 3 and table 4, the self-sensing displacement of piezoelectric ceramic stack is shown in table 5.

Table 5. Self-sensing displacement and first converse displacement.

Driving voltage（V）	10*10*30.4		5*5*121.6	
	first converse Displacement（μm）	Self-sensing displacement（μm）	first converse displacement（μm）	Self-sensing displacement（μm）
0	0.013	0	0.0276	0
10	1.477	0.66	5.863	3.45
20	3.453	1.49	14.017	6.76
30	5.683	2.23	20.863	10.08
40	8.113	2.71	29.087	13.35
50	10.333	3.53	37.040	16.96
60	12.727	4.44	45.087	20.21
70	15.247	5.19	52.963	23.54
80	17.513	5.94	60.890	26.93
90	18.310	6.93	72.747	31.23
100	21.773	7.68	80.653	34.53

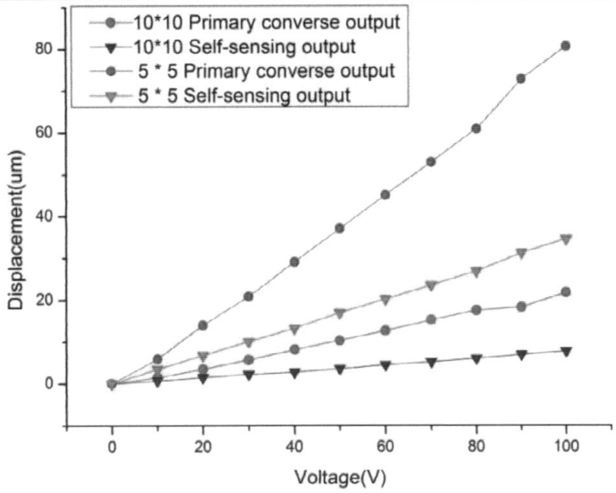

Figure 8. Comparison diagram.

Although there is a large error between the self-sensing displacement and first converse displacement of piezoelectric ceramic stack, there is still a certain proportional relationship. The self-sensing displacement of 10*10*30.4 type ceramic stack is about 0.4 times of an inverse actual measurement of displacement. The self-sensing displacement of 5*5*121.6 type ceramic stack is about 0.5 times of an inverse actual measurement of displacement. According to the relationship between output and input, the input can be set between 0V~80V to ensure a linear relationship.

3. Model design and experimental verification

3.1. Model design

Based on the requirements of cell injection and the driving characteristics of different fluids, the multiple piezoelectric effect characteristics of piezoelectric ceramic stack were used to make small changes in the content of the syringe filled with medicine liquid for microinjection. Using the above theory and experiment, the microinjection device can complete the displacement and injection volume during the injection process. It can perceive the actual displacement and injection volume, and realize the positioning and quantitative control of microinjection to ensure the reliable and stable output of the liquid[10].

Through the analysis of functional structure and the process of microinjection, the corresponding structures are designed respectively. As shown in figure 8, it consists of four parts: 10*10*30.4 motion mechanism (positioning control), 5*5*121.6 motion mechanism (quantitative control), injection motion mechanism and peripheral detection circuit.

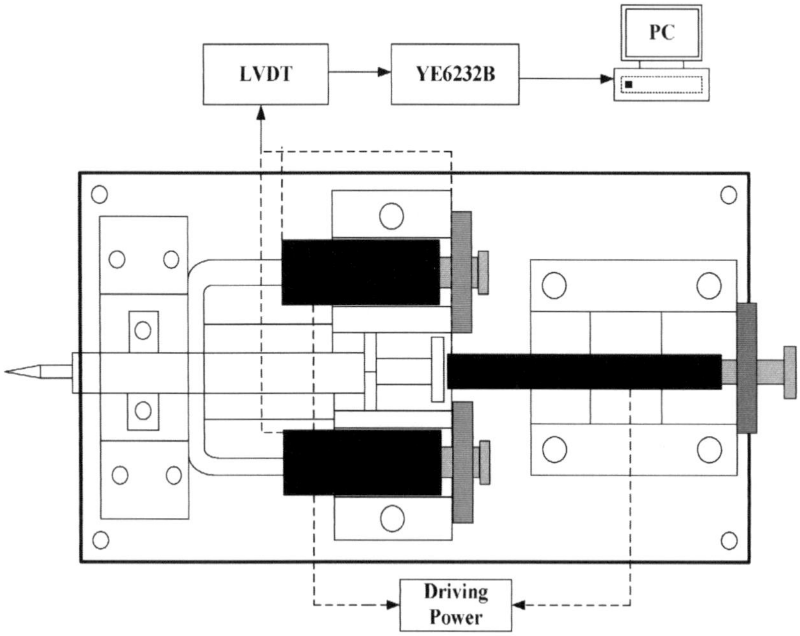

Figure 9. Structure diagram of piezoelectric microinjection device.

*3.1.1. 5*5*121.6 series motion mechanism*

This structure is used to push the syringe piston and inject the liquid into the object. In figure 10, base 1 is fixed on the platform with fixed screw 2, and slot 3 is open on base 1, and 5*5 stack is placed into straight slot to prevents the body from moving to right and left. The pretightening screw 4 is placed at the end of straight slot to secure one end of the stack and the other end is freely retractable.

*3.1.2. 10*10*30.4 series motion mechanism*

This structure is used to push the entire syringe to move the micro needle out of the cell, which is similar to the 5*5 motion mechanism in the structural design. In figure 11, two ceramic stacks are used to make the entire syringe move more accurately, base 5 is fixed on the platform, and there are two grooves 6, which are used to place the 10*10 stack, One end of the ceramic stack is fixed with pretightening screw 7, and the other end is connected with a u-shaped block 8. When the ceramic stack is energized, the u-shaped block is moved, and the u-shaped block is used to drive the syringe motion system connected in front.

IOP Conf. Series: Journal of Physics: Conf. Series **1209** (2019) 012001 doi:10.1088/1742-6596/1209/1/012001

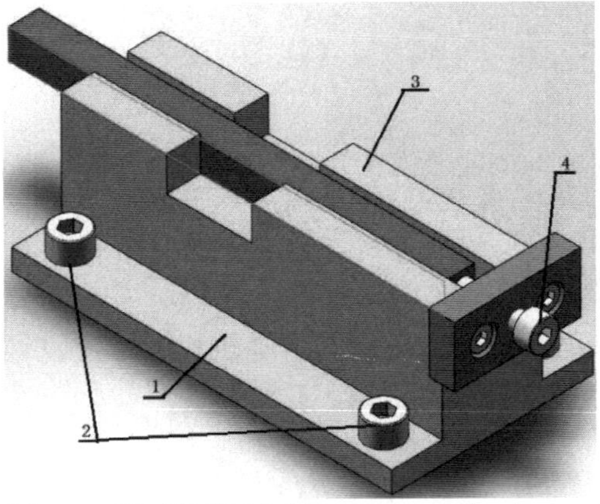

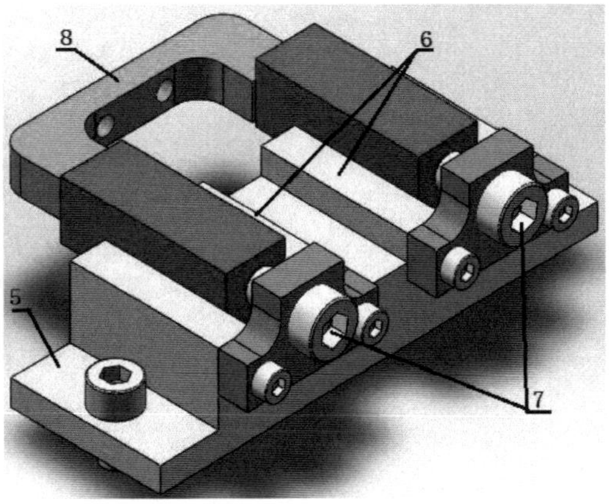

Figure 10. 5*5*121.6 Driving mechanism. **Figure 11.** 10 10*10*30.4 Driving mechanism.

3.1.3. Injection motion mechanism

This structure is mainly used to realize the movement and fixation:

(1) Movement: U-shaped block 8, the needle tube 9, the suction plate 10, and the linear guide 11 are fixed together by fastening screws, and 11 and 12 constitute a frictionless track. When the 10*10 ceramic stack is energized, a telescopic deformation occurs, which causes the needle to move on the frictionless track to determine the position of the injection.

(2) Clamping: When the injection position is determined, the magnet in the base is energized, and the adsorption plate 10 is fixed by the generated suction. The position of the syringe does not change. When the 5*5 stack is energized, the deformation causes the piston to move forward, and then the liquid is squeezed out and injected into the cells to complete the microinjection.

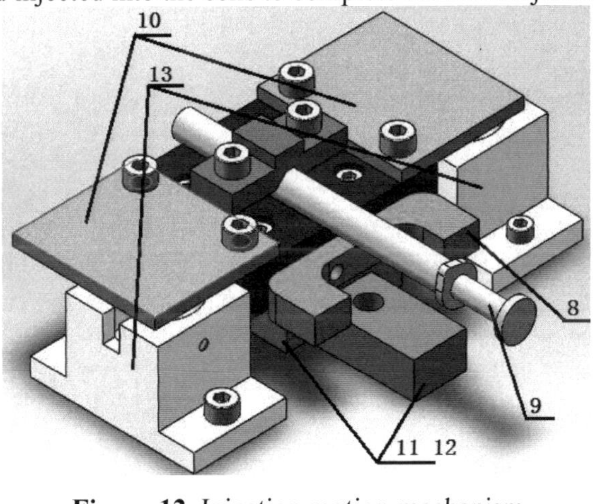

Figure 12. Injection motion mechanism

3.2. Experimental verification

Figure 13 is the physical structure diagram of the self-sensing injection device. Firstly, 10*10 stack was preloaded, and then voltage was applied to the ceramics, resulting in micro displacement to drive the whole syringe movement, and there will be a gap between the 5*5 stack and the plunger of the syringe. After the ceramic is pre-tightened again through the screw, the liquid injected will reach the critical state at the needle. Finally, a voltage was applied to the 5*5 stack, and the resulting displacement pushed the piston to move slightly, injecting the liquid into the cells. In the whole process, the ceramic stack perceiving actual output displacement through secondary direct piezoelectric effect.

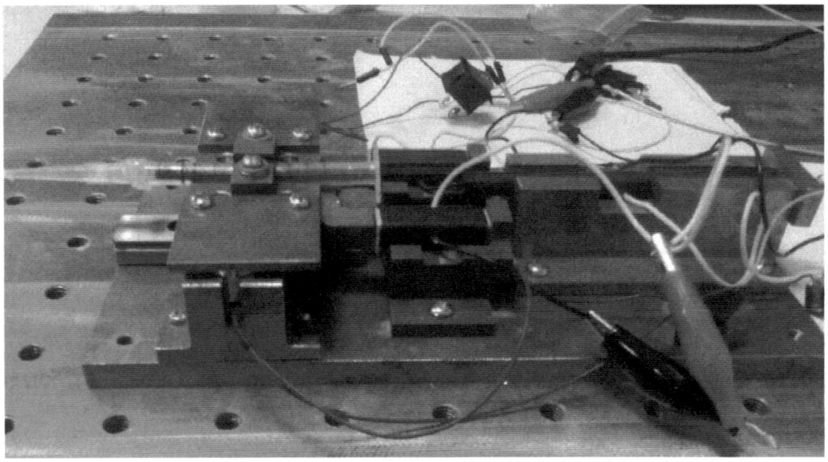

Figure 13. Physical structure of piezoelectric microinjection device.

3.2.1. Microinjection flow generation experiment

Applying 0V~80V drive voltage to the 10*10 ceramic motion mechanism, the resulting micro displacement is shown in table 6.

Table 6. Position measurement.

Driving voltage（V）	Experimental displacement 1(μm)	Experimental displacement 2(μm)	Average displacement (μm)
0	-0.15	-0.47	0.31
10	-1.17	-1.51	1.34
20	-2.13	-2.63	2.38
30	-3.46	-3.96	3.71
40	-4.92	-5.58	5.25
50	-6.64	-7.34	6.99
60	-8.39	-9.02	8.705
70	-10.13	-10.67	10.4
80	-11.91	-12.26	12.085

After positioning, electrify the magnet of the motion mechanism, fix the adsorption plate, and apply 0V~80V drive voltage to the 5*5 ceramic stack motion mechanism. In the experiment, the thinnest conventional injection needle was selected. When the drive voltage reached 40V, small droplets would appear in the tip, and the droplets would increase with the increasing voltage.

3.2.2. Microinjection flow measurement experiment

When the microinjection system is fixed, the voltage is applied from 40V. Under the same voltage, the same operation is repeated 5 times. The scale is used to measure the total liquid outflow for 5 times, and then the average value is obtained. Then, the drive voltage is transformed, 10V is a benchmark, and each drive voltage is measured 5 times until 80V, as shown in figure 14 and 15.

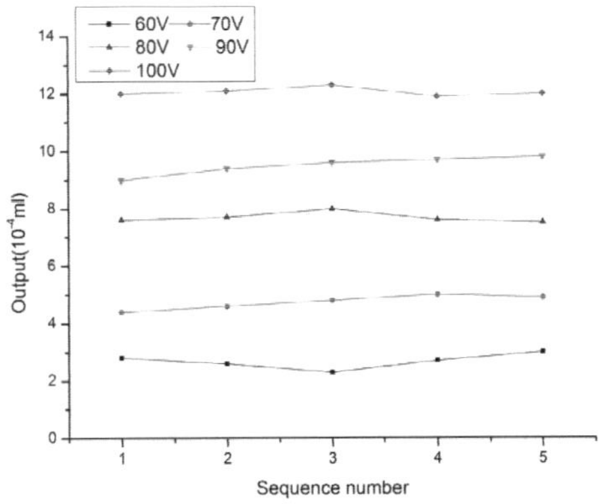

Figure 14. Flow rate is measured under different voltage.

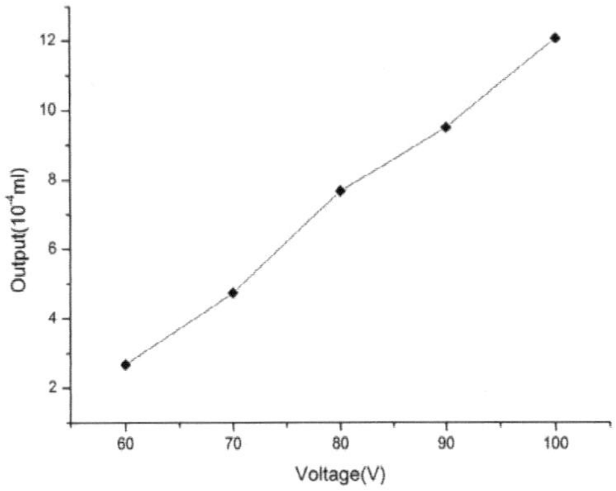

Figure 15. Voltage and flow diagram.

3.2.3. Microinjection flow perception experiment
The measurement results of microinjection flow perception are shown in table 7. According to the analysis of the table, the perceived displacement is 0.4 times of the actual displacement of the positioning mechanism, and the injection mechanism is 0.5 times, the output micro displacement has a good linear relationship with the applied voltage, which proves the correctness of the theoretical analysis and the rationality of the experimental setup.

Table 7. Self-sensing measurements.

Driving voltage(V)	Self-sensing displacement of 10*10 (μm)	Self-sensing displacement of 5*5 (μm)
0	0.075	0
10	0.385	0.54
20	1.5	1.53
30	1.235	2.56
40	1.565	3.63
50	2.385	4.62
60	2.995	5.64
70	3.59	6.62
80	4.1	7.61

The driving voltage can be applied to control the displacement and injection amount by the first converse piezoelectric effect, the actual displacement and injection amount are perceived by the secondary direct piezoelectric effect, and achieving the quantitative control of microinjection.

References

[1] Chen GH, Yan XG, Cai JH and Guo II 2018 Hysteresis nonlinear modeling and compensation of piezoelectric ceramic sensors in micro measurement systems *Measurement Science and Technology*

[2] ZHANG Zhong-hua, SUN Bao-yuan and QIAN Min 2009 Influence of Multiple Piezoelectric Effect on Static Performance of Piezoelectric Sensor *Piezoelectrics and Acousto-optics*

[3] WEI Yanbo, SHI Liping, WEI Xiwen and HUANG Jie 2014 Research on Micro-flow Self-sensing Actuators based on Piezoelectric Ceramic Stack *Journal of Harbin Institute of Technology*

[4] SHI Liping, HUANG Jie, WEI Xiwen and WEI Yanbo 2014 Investigation on Multi-piezoelectric Effects from the First Positive Piezoelectric Effect *Key Engineering Materials*

[5] Payri R, Gimeno J, Mata C and Viera A 2017 Rate of injection measurements of a direct-acting piezoelectric injector for different operating temperatures *Energy Conversion and Management*
[6] SUN Kang and ZHANG Fuxue 1984 *Beijing Piezoelectricity* (National Defense Industry Press)
[7] ZHANG Fu-xue and WANG Li-kun 2001 *Beijing Modern piezoelectricity* (Science Press)
[8] Shi Liping, Sun Lining and Rong Weibin 2010 Application of the first inverse and the secondary direct piezoelectric effect to self-sensing actuators *Nanotechnology and precision engineering*
[9] SHI Li-ping, WEI Xi-wen and WEI Yan-bo 2012 Comparative Analysis on Piezoelectric Effect and Electromagnetic Effect for the Further Study of Multi-piezoelectric Effect *International Conference on Measurement, Information and Control(MIC)*
[10] Xiao Xianqiang, Li Xinxin, et al 2005 Research on the device of cell medicine microinjection based on piezoelectric actuator *Machinery Design & Manufacture*

Acknowledgments
This work was supported by the National Special Fund for Forest Scientific Research in the Public Welfare and Heilongjiang Higher Education Project.

Design and research of supplying power for spinning emitter in needleless electrospinning with non-metallic rotating shaft

J Liu[1,4,*], Y B Liu[2,3], X M Jiang[1], X Luo[3] and M Zeng[3]

[1] Tianjin Key Laboratory of Modern Technology & Equipment, Tianjin Polytechnic University, Tianjin, 300387, China;
[2] School of Textiles Science and Engineering, Wuhan Textile University, Wuhan 430200, China;
[3] Key Laboratory of Advanced Textile Composites of Ministry of Education, Tianjin Polytechnic University, Tianjin, 300387, China;
[4] National Experimental Teaching Demonstrating Center of Engineering Training,Tianjin Polytechnic University, Tianjin, 300387, China

[*]Correspondence: liujian3286@126.com; Tel.: 13512912151

Abstract. Electrospinning is a widely used technique for nanofiber preparation at large scale that is mainly divided into two categories, namely, multi-needle and needleless categories. Among them, needleless electrospinning technologies are being in-depth researched and developed, because there is no needle blockage, electrostatic interference between the needles, and the yield is high. At present, the needless electrospinning devices with rotating shaft are applied widely. In order to supply power for the spinneret emitter conveniently, the metal shaft is used mostly and add electricity for it by brush. However, it not only weakens the utilization of electric energy, but also affects the intensity and the distribution of the electric field, and then affects the quality of the electrospinning. In this paper, a design of supplying power for spinneret emitter with a non-metallic rotating shaft was proposed, and the diameter formula of non-metallic rotating shaft under different electrospinning conditions was deduced. The intensity and distribution of the electric field of this design scheme is researched based on a fractal electrospinning head by the finite element analysis software COMSOL Multiphisics 5.0. At last, the electrospinning experiments were carried out by the electrospinning devices with non-metallic rotating shafts to prove the optimization effects.

1. Intorduction

Electrospinning is an effective technique to produce nanofiber using electric field force to stretch the spinning jets from the polymer solution [1]. Currently the electrospinning technique which enables mass production of nanofibers has been classified as two categories, i.e., multi-needle type (with capillary needles) [2, 3] and needleless type [4–13]. Multi-needle electrospinning, in spite of the advantage of higher productivity than single needle, its disadvantage such as spinning channel clogging and difficulty in cleaning, especially the "End effect" has severely hindered its industrialization due to its generation of non-uniformity in electric field intensity [14]. Needleless electrospinning technique is a spinning method using high voltage electric field to form jets on the free liquid surface. There are not the problems such as needle blockage and mutual interference between needles, which can greatly improve the yield of electrospinning. Therefore, a lot of needleless electrospinning devices has been researched and manufactured in recent years, such as roller and wire commercialized by Elmarco with the brand name Nanospider [5, 6], as well as rotating disk [7], spiral coil [8], magnetic suspension [12], and so on, have been used to produce nanofibers at large scale. Among them, the needleless electrospinning devices with rotating spinning emitters are used widely.

The rotating shaft almost is metallic material in these needleless electrospinning devices. It is a common method to use the brush to supply power for the metal rotating shaft, and then to make charges for the spinning emitters. However, the induced electric field of the metal rotating shaft affects the one of the spinning emitters, and then affects the spinning effect. Meanwhile, it will cause waste of energy. Therefore, it is of great significance to use non-metallic rotating shaft for the needleless electrospinning devices. However, it is a problem that how to supply power for the spinning emitters with non-metallic rotating shaft and minimize the influence of the electric field on the spinning emitters.

Aiming at the above problems, a method supplying power for the the spinning emitters in needleless electrospinning with non-metallic rotating shaft was proposed, and the diameter formula of the non-metallic rotating shaft under different spinning conditions was deduced. Then a kind of electrospinning head based on the fractal theory with non-metallic rotating shaft was designed, and its electric field distribution and power utilization ratio were researched by the finite element analysis software COMSOL Multiphisics 5.0, at last, the optimization effects were proved by the electrospinning experimentals.

2. Design non-metallic rotating shaft

A kind of non-metallic rotating shaft in needleless electrospinning devices need to be designed to minimize the influence of the electric field on the spinning emitters, reduce energy consumption, and prevent wire winding.

2.1. Structure design of non-metallic rotating shaft

As shown in Figure 1, the non-metallic rotating shaft is composed of central shaft, supported sleeve, end covers, jump ring, bearings, powered ring, housings, metal wire Ⅰ, and metal wire Ⅱ. The supported sleeve and the end covers on both sides are collectively called "roller". In order to minimize the influence of the electric field, reduce the weight, and be easily manufactured, the material of the central shaft and the roller can choose the plastic with high corrosion resistance, high strength, and high insulation. There are N holes on the supported sleeve, and the metal wire Ⅰ passes through and links with the spinning emitter. If the spinning emitter is discontinuous single type (such as rotating disk), N is the number of the discontinuous single type. If the spinning emitter is continuous type (such as spiral coil), N equals one, and the hole should be close to the right end cover. There are two holes in the central shaft and on both sides of the right end cover, respectively named hole a and hole b, and they intersects with central hole. The jump ring contacts with shoulder c. The right bearing contacts with the jump ring. The powered ring is a metal thin slice and located in the bearing bore of the right housing and contacts with the right bearing.

The key problem of supplying power to the spinning emitter with non-metallic rotating shaft is the link method of the metal wires. The specific link method is as follows: one end of the metal wire Ⅰ is connected with the spinning emitter, and the other end is penetrated in the hole a, and penetrated out from the hole b, then connected with the jump ring; one end of the wire Ⅱ is connected with the powered ring, and the other end is penetrated through the hole in the right housing, then the right bearing is mounted into the bearing bore and presses the powered ring. In this way, the positive electrode of the direct high voltage power supply is connected with the wire Ⅱ, and the spinning emitter also is charged through the powered ring, right bearing, jump ring and wire Ⅰ.

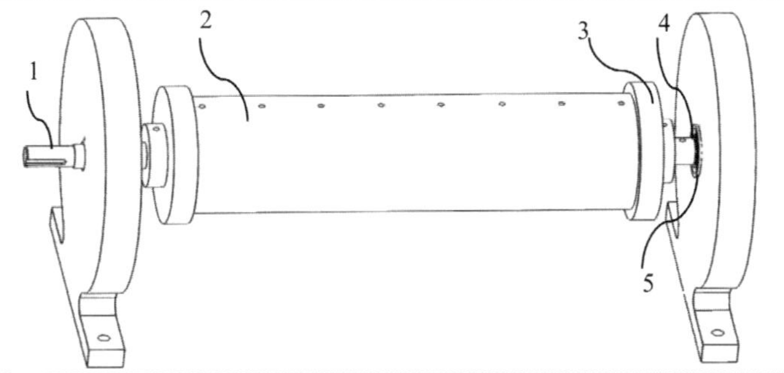

(a)

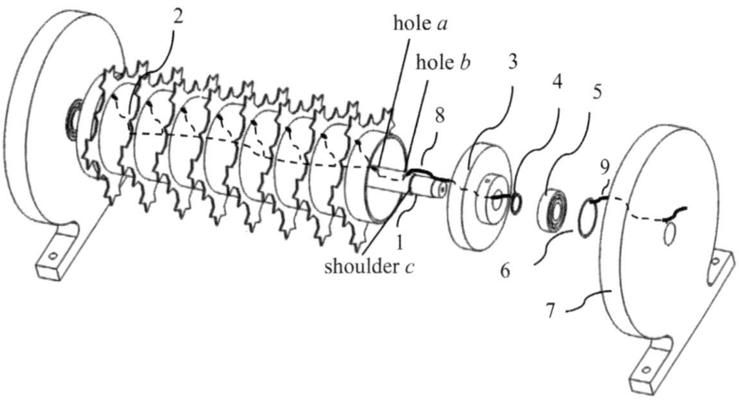

(b)

Figure 1. Structure of the non-metallic rotating shaft: (a) The assembly drawing; (b) The local exploded drawing; 1.Central shaft, 2.Supported sleeve, 3.End cover, 4.Jump ring, 5.Bearing, 6.Powered ring, 7.Housing, 8.Metal wire Ⅰ 9.Metal wire Ⅱ.

2.2. Parameters calculation of the non-metallic rotating shaft

The key parameters of the non-metallic rotating shaft mainly includes the dimensions of the roller (such as outside diameter *D*, wall thickness a and the length *L*) and the minimum diameter *d* of the central shaft. Among them, *D* and L mainly depend on the shape, layout and inside diameter of the spinning emitter, and they can be used as known quantities. The wall thickness a can be determined according to the strength of the material of the supported sleeve. Then the total mass *M* of the roller and the spinning emitter is equivalent to the known quantity.

The force analysis of the non-metallic rotating shaft is shown in Figure 2. The minimum diameter *d* of the central shaft is calculated according to strength checking based on allowable bending stress. The main steps are as follows.

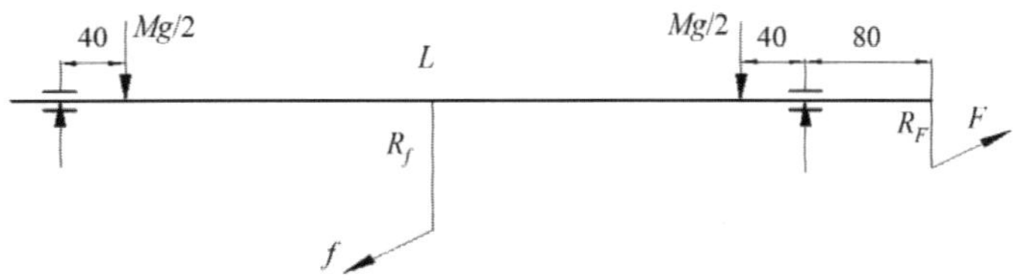

Figure 2. Force analysis diagram of the non-metallic rotating shaft.

(1) Calculate the viscous resistance f acting on the spinning emitter

τ_f is the shear stress of the non-metallic rotating shaft acted by spinning polymer solution. And

$$\tau_f = K\left(\frac{dv}{dr}\right)^n = K\overline{\gamma}^n \qquad (n<1)$$

where K and n are the consistency index and non-Newtonian index of the spinning polymer solution respectively, and they are constant. The viscous resistance f can be calculated by Equation (1).

$$f = A \cdot \tau_f , \tag{1}$$

Among it, A is the contacting area between the spinning emitter and the spinning polymer solution.

(2) Calculate the torque T acting on the non-metallic rotating shaft

Measure the distance R_f between the highest point of the spinning emitter and the central axis. And then $T = f \cdot R_f = A\tau_f \cdot R_f$.

(3) Calculate the bending moment M acting on the non-metallic rotating shaft

$M = (Mg/2) \cdot (L/2 + 40)$

(4) Calculate the equivalent bending moment M_e of the risk section

$M_e = \left[M^2 + (\alpha T)^2\right]^{1/2}$

As the speed of the electrospinning process is generally low, so the reduction coefficient α is selected as 0.3. And then

$$M_e = \left[M^2 + (\alpha T)^2\right]^{1/2} = \left[(Mg/2)^2 \cdot (L/2 + 40)^2 + (0.3 fR_f)^2\right]^{1/2} \tag{2}$$

(5) Calculate the minimum diameter d of the non-metallic rotating shaft

According to the ending-torsion strength checking method [16], there is Equation (3).

$$d \geq 21.68 \times \left(\frac{M_e}{0.1 \cdot \sigma_{-1}}\right)^{1/3} \tag{3}$$

where σ_{-1} is the allowable bending stress of the non-metallic rotating shaft, and can be retrieved from reference [16].

3. Example

The electrospinning head (illustrated in Figure 3) is designed by our research group based on the fractal theory [17]. We take it as an example to determine the specific dimensions of the non-metallic rotating shaft according to the above contents.

The parameters of the fractal slice are as follows: the inside diameter is 69mm, the thickness is 2mm, and the distance R_f between the highest point of the fractal slice and the central axis is 110mm, there are five fractal slices at least, and the distance between them is 40mm. So then the outside diameter of the roller D is also 69mm, and its length L is 272mm. The wall thickness is chosen as 10mm. The weight M of the electrospinning head (including the roller and the fractal slices) is 6kg.

As shown in Figure 4, the dotted line indicates the liquid level. A_1 is the maximum connected area between the fractal slice and polymer solution. According to the specific sizes of the fractal slice in reference [17], A_1 is calculated and it is 550.98mm^2. So the total connected area A of the spinning emitters with five fractal slices is $A = 2\times5\times A_1 = 10\times550.98 = 5509.8\text{mm}^2 \approx 0.0055\text{m}^2$.

When calculating the diameter of the non-metallic rotating shaft, the polymer solution with higher concentration and suitable for electrospinning should be chosen. For example, we prepared polyvinyl alcohol solution with a concentration of 25%. The measured consistency index K is 63.59, and the measured non-Newtonian index is 0.9645 by Malvin rotation rheometer.

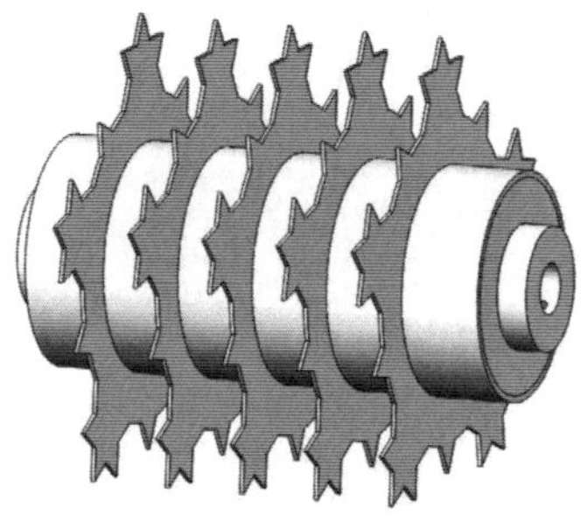

Figure 3. Electrospinning head based on the fractal theory.

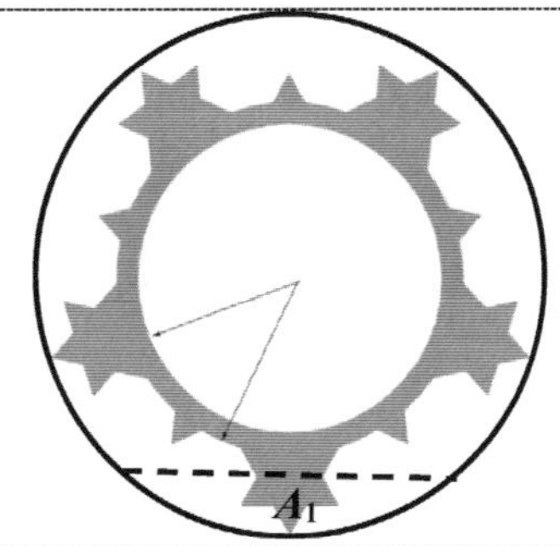

Figure 4. The area of the fractal slice in the liquid.

Based on the experience of electrospinning experiment, the suitable speed is 300r/min, so the angular velocity ω=2π×300/60=31.4rad/s. According to Equation (1), the viscous resistance f was calculated.

$$f = A\cdot\tau_f = AK\left(\frac{dv}{dr}\right)^n = AK\left(\int_{79}^{110}\varpi dr\right)^n = 0.0055\times 257.2\times\left(\int_{79}^{110}31.4dr\right)^{0.7125} = 178.77N$$

The equivalent bending moment M_e was obtained by Equation (2), and is approximately equal to 7.85Nm.

The material of the central shaft is nylon, and its allowable bending stress σ_{-1} is determined 70MPa by checking conference [16]. And then the minimum diameter of the non-metallic rotating shaft was worked out by Equation (3).

$$d = 21.68\times\left(\frac{M_e}{0.1\cdot\sigma_{-1}}\right)^{1/3} = 21.68\times\left(\frac{7.85}{0.1\times 70}\right)^{1/3} \approx 22.5mm$$

4. Electric field simulation

Taking the electrospinning head with five fractal slices and metal rotating shaft or non-metallic rotating shaft as the research object, the electric field simulation was carried out by the finite element analysis software COMSOL Multiphisics 5.0.

The electrostatic field simulation follows Poisson equation in the finite element analysis software COMSOL Multiphisics 5.0, as follows:

$$-\nabla d\varepsilon_0\varepsilon_r\nabla V = d\rho$$

where ε_0 and ε_r are vacuum relative permittivity and dielectric relative permittivity respectively, V is potential energy, and ρ is space charge density. In this paper, dρ is zero. The value of ε_r is different according to different material. The receiving plate and the spinning emitter are metal (steel was chosen in this paper), and their ε_r are 1.5. The material of the roller and the non-metallic rotating shaft is nylon, and their ε_r are 2.55. In addition, the corresponding electric strength and electric displacement vector can be obtained according to the equations $E=-\nabla V$ and $D=\varepsilon_0\varepsilon_r E$ (E is the electric strength and D is the electric flux density).

In the simulation, the four boundaries of the air model are set to zero charge and symmetry condition. In this way, the limited model can be considered as infinitely far. And then the bottom of the receiving plate is grounded. The spinning emitter is supplied the required voltage. And the other boundary conditions are continuous. Table 1 shows the parameters of the models and the simulation.

Table 1. The parameters of the models and the simulation.

Voltage (kV)	Receiving plate (mm)	Receiving distance (mm)
30	600 × 500 × 1	300

IOP Conf. Series: Journal of Physics: Conf. Series **1209** (2019) 012002 doi:10.1088/1742-6596/1209/1/012002

The material of the spinning emitter (i.e. fractal slices) and the roller is steel and nylon respectively in Figure 5. The material of the rotating shaft in Figure 5(a) and Figure 5(b) is steel and nylon respectively. The color of the two ends of the metallic rotating shaft is bright in the electric field simulation, as shown in Figure 5(a). It shows that a higher electric field is also produce on the metallic rotating shaft. And it affects the electric field of the spinning emitter, but also has the risk of spark-over with the external equipments, and then affects the normal operation of the spinning. Figure 5(c) illustrates the specific values of the electric strength and distribution of the electric field. It shows that the maximum electric strength with the metallic rotating shaft is 1.20×10^6 V/m, and the average value is 9.48×10^5 V/m; and the maximum electric strength with the non-metallic rotating shaft is 1.33×10^6 V/m, and the average value is 1.03×10^6 V/m(i.e. increased by 8.65% compared to the metallic rotating shaft).

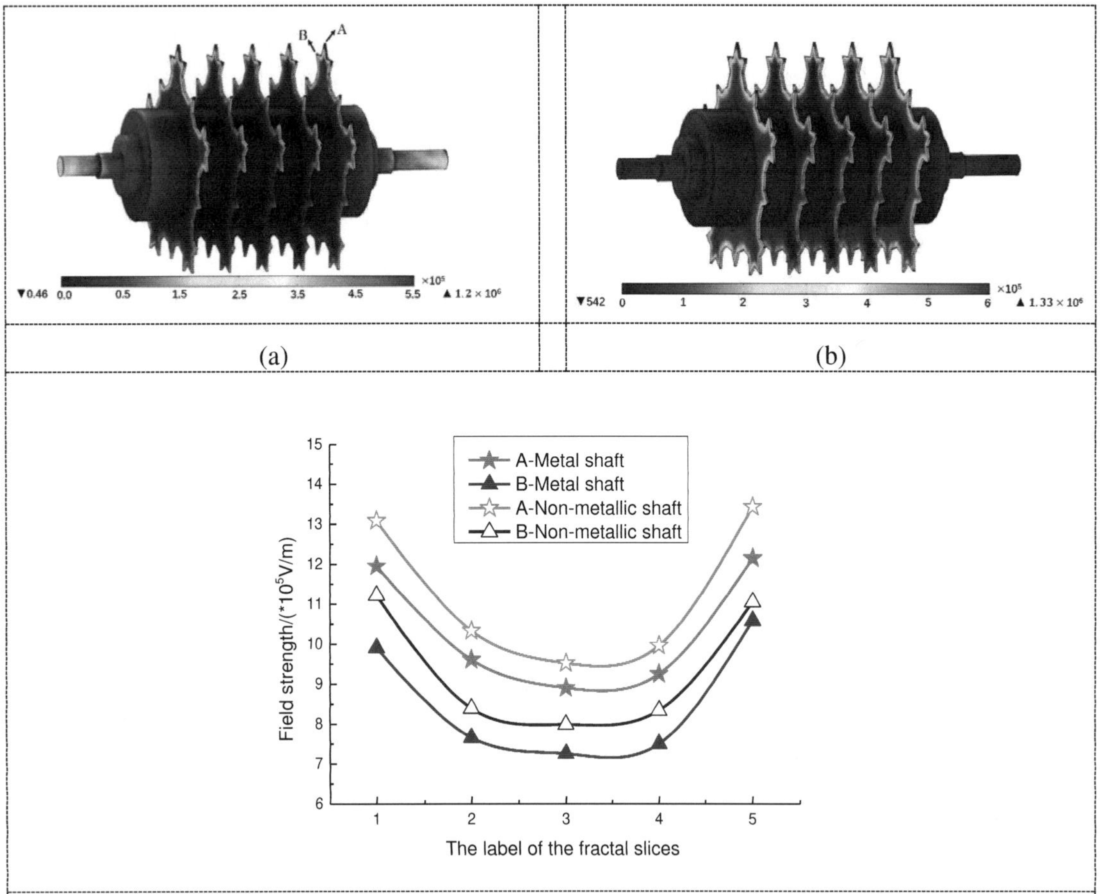

(a) (b)

(c)

Figure 5. Simulation of the electric field with different rotating shaft: (a) The electrospinning head with metal (steel) shaft; (b) The electrospinning head with non-metallic (nylon) shaft; (c) The electric field strength of the spinning heads with different rotating shafts.

5. Electrospinning experimental

5.1. Spinning solution

The degree of polymerization, hydrolysis, and the relative molecular mass of the polyvinyl alcohol (PVA) (Shanghai Petrochemical Co, Ltd., China) in the experiment was 1700, 99% and 74800 respectively. The distilled water was self-made. An appropriate amount of PVA and distilled water were weighed and mixed according to the concentration 15% of the solution. A magnetic rotor was placed in the solution. Subsequently, the solution was transferred to a thermostatic bath and stirred at 80°C until the solution was clear and transparent [18]. Finally, the experimental polyvinyl alcohol solution with a concentration of 15% was prepared.

5.2. Spinning devices

The experiment was performed by the self-made electrospinning device with five fractal slices. Table 2 shows the types and sources of the related instruments.

Table 2. Experimental instruments.

Instruments	Type	Source
Spinning device	Five fractal slices	Self-made
DC high-voltage power supply	DW-P/N603	Tianjin Dongwen High Voltage Power Supply, Ltd.
Metal halide lamp	70W	Xincheng Lighting, Ltd.
Motor agitator	DF-101S	Gongyi Yuhua Instrument Co., Ltd.
Thermostat water bath	HH-4	Kexi Instrument, Ltd.
Camera	Sony DSC-TX9	Sony
Electron microscope	TM3030	Techcomp (China) Ltd

5.3. Experimental parameters

The depth of the fractal slice in the polymer solution was 15mm. The rotating speed of the spinning head was 300r/min. And the receiving distance was 300mm. A DC power of 25 kV was supplied by the positive voltage to the wire of the non-metallic rotating shaft, while a negative voltage of 5kV was applied to the receiving electrode. The spinning time was 10 min. The spinning process was shown in Figure 6.

5.4. Experimental results

The scanning electron microscopy (SEM) image of the nanofiber film prepared by the electrospinning experiment was shown in Figure 7. And then the diameters of one hundred fibers was measured by the software named Image-Pro Plus6.0. Finally, the average diameter and CV value of the fibers were calculated, and they were 324nm and 17.3% respectively.

Figure 6. Electrospinning experiment with non-metallic rotating shaft.

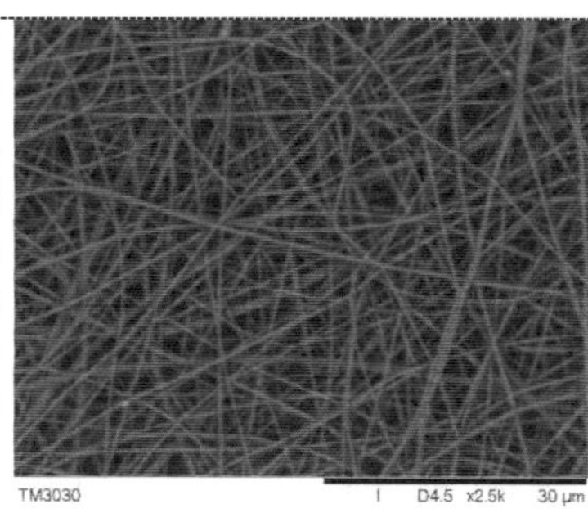

Figure 7. SEM of the prepared fibers.

6. Conclusions

In this paper, a design scheme of supplying power for spinning emitter with non-metallic rotating shaft in the needleless electrospinning technique was proposed. And the diameter formula of the non-metallic rotating shaft under different spinning conditions was deduced. Then a kind of electrospinning head based on the fractal theory with non-metallic rotating shaft was designed, and as an example, its electric field distribution and power utilization ratio were researched by the finite element analysis software COMSOL Multiphisics 5.0. The study found that the average value of the field strength of the spinning emitter with non-metallic rotating shaft is increased by 8.65% compared with the one with metal rotating shaft. And finally, electrospinning experiment was carried out by the self-made needleless electrospinning device base on the fractal theory, and the average diameter of the prepared electrospun membrane is 324nm， and the CV value is 17.3%. Therefore the energy consumption is reduced by using the non-metallic rotating shaft in the needleless electrospinning devices, at the same time, the spinning quality is accepted. What's more, the linking method of the wires is simple and prevents the winding problem.

References

[1] Jin S X, Xin B J and Zheng Y H 2018 Overview on mass production of electrospun nanofibers *J. Textile Res.* **39** 175-80
[2] Theron S A, Yarin A L, Zussman E and Kroll E 2005 Multiple jets in electrospinning: experiment and modeling *Polymer* **46** 2889-99
[3] Alessio V, Riccardo A C and Giorgio M 2009 Experimental investigations on the multi-jet electrospinning proces *J. Mater. Process. Technol.* **209** 5178-85
[4] Niu H T, Wang X G and Lin T 2011 Needleless electrospinning: developments and performances (Intech: Rijeka, Croatia) pp 17-36
[5] Oldrich J, Filip S, David L, Vaclav K, Lenka M and Jiri C 2005 A method of nanofibres production from a polymer solution using electrostatic spinning and a device for carrying out the method *WIPO Patent* 2005/024101
[6] Huang X P, Wu D Z, Zhu Y Y and Sun D H 2007 Needleless electrospinning of multiple nanofibers *Proceedings of the 7th IEEE International Conference on Nanotechnology* 823-26
[7] NIU H T, LIN T and WANG X G 2009 Needleless electrospinning. i. a comparison of cylinder and disk nozzles *J. Appl. Polym. Sci.* **114** 3524-30
[8] Wang X, Niu H T, Wang X G and Lin T 2012 Needleless electrospinning of uniform nanofibers using spiral coil spinnerets *J. Nanomater* **5920** 1875-90
[9] Wang X, Niu H T, Lin T and Wang X G 2009 Needleless electrospinning of nanofibers with a conical wire coil *Polym. Eng. Sci.* **49** 1582-86
[10] Lu B G, Wang Y J, Liu Y X, Duan H G, Zhou J Y, Zhang Z X, Wang Y Q, Li X D, Wang W, and Lan W 2010 Superhigh-throughput needless electrospinning using a rotary cone as spinneret *Small* **6** 1612-16
[11] Jiang G J, Zhang S and Qin X H 2013 High throughput of quality nanofibers via one stepped pyramid-shaped spinneret *Mater. Lett.* **106** 56-58
[12] Yarin A L and Zussman E 2004 Upward needleless electrospinning of multiple nanofibers *Polymer* **45** 2977-80
[13] He J H, Liu Y, Xu L, Yu J Y and Sun G 2008 Biomimic fabrication of electrospun nanofibers with high-throughput *Chaos Soliton. Fract.* **37** 643-51
[14] Aaron J R and Richard C F 1993 Scale-up of electrospray atomization using linear arrays of taylor cones *Rev. Sci. Instrum.* **64** 683-86
[15] Cheng D X 2010 Mechanical design manual: shaft and its connection (Beijing: Chemical industry press) pp. 20-21
[16] Cheng D X 2010 Mechanical design manual: common engineering materials (Beijing: Chemical industry press) pp. 20-21
[17] Yang W X, Liu Y B, Zhang L G, Cao H, Wang Y and Yao J B 2016 Optimal spinneret layout in

von koch curves of fractal theory based needleless electrospinning process. *Aip Adv.* **6** 065223
[18] Araujo E S, Nascimento M L F and Oliveira H P D 2016 Electrospinning of polymeric fibres: an unconventional view on the influence of surface tension on fibre diameter. *Fibres Text. East. Eur.* **1** 22-29

Acknowledgments
The authors gratefully acknowledge financial support from the National Natural Science Foundation of China (No. 51373121).

The vertical placement ultra - thin flameproof baffle of the fuze

C C Sun [1,2], H B Liu [1,2] and X Zhao[1,2]

[1]Xi'an Institute of Electromechanical Information Technology, Xi'an 710065, China
[2] Science and Technology on Electromechanical Dynamic Control Laboratory, Xi'an Shaanxi, 710065, China

002sc0603@sina.com

Abstract. In order to meet the operational requirements of a new type of intelligent small penetrating submunition fuze with large overload of target base, a new type of ultra-thin vertical fuze flameproof baffle is proposed. Vertical arrangement of fuze ultra-thin flame-proof board uses the three layers of different vertical composite plate structure of metal materials, the overall size for the 1.6mm thickness, the length of vertical is 20mm, upper and lower layer 0.3mm thickness of tungsten-copper alloy plate, respectively, the middle layer of 304 stainless steel. The simulation and test results show that the ultra-thin vertical fuze baffle can resist the penetration overload of 120,000g, and guarantee the safety of 12m dropping and explosion isolation.

Key words. Submunition fuze; Explosion isolation mechanism; Vertical layout; Ultra-thin type flameproof plate

1. The introduction

The development direction of new intelligent miniaturized ammunition has put forward a new military demand for new type of small penetration submunition medicine. Fuze, therefore, in addition to the original function, but also increase the wireless communication module, more sensors and actuators, functional components, such as the miniaturization of the fuze safety system design put forward higher requirements.

Requirements for fuze in fuze limited volume to accommodate more sensors, actuators to perception, ballistic, target detection and processing more environmental information, such as complete more complex instructions and ensure the reliability, meet the demand of the expanded function and performance improvement of ammunition, but also increased the reliable firing rate of dual redundant detonating sequence. The new generation of intelligent fuze system is far more complex than traditional fuze system, but the fuze volume is limited, fuze must realize miniaturization to meet the increasing functional requirements. The miniaturization of fuze security system can make room for functional components, which is one of the main ways of fuze miniaturization.

At present domestic security system miniaturization has been studied for more than 20 years, the second generation of miniaturization of latest security system in our country, in such aspects as performance, volume has same level with the third generation of miniaturization of security systems[1]. The United States, Russia, France, Israel and other countries have developed the latest fourth-generation miniaturized security system, which is smaller and safer[2].In 2010, the United States announced the development plan of its miniaturized fuze security system, requiring the size of the entire safety system to be smaller than0.44in3 (about 7.2cm3)[3], the average size of each direction is less than 2 cm. The flameproof plate is an important part of the safety system. In order to adapt to the new type of miniaturized fuze safety system, the size of the flameproof plate structure also needs to be miniaturized.

Because the fuze function of the new type of small penetration submunition is increased and the space is limited, it is necessary to design a double-path explosion isolation mechanism to increase the double-path redundant transmission sequence. Due to the small big overload of penetration submunition to target base to remove flame-proof and trigger delay ignition circuit based on the touch targets at the same time, thus flame-proof institutions need to be used along the elastic axis vertical arrangement to meet requirements.

In order to meet the operational requirements of a new type of intelligent small penetrating submunition fuze with large overloading of the target base, a new type of ultra-thin vertical fuze baffle plate is proposed.

2. Background technology

2.1. Small-sized single-way sliding barrel explosion isolation mechanism arranged vertically by submunition fuse

Figure 1 shows the vertical arrangement of a small penetration bullet fuze with a single slide cylinder type explosion isolation mechanism, made of 45 steel. Because of the small target medicine rely on penetration bullet base big overload lifting and isolation, so the flame-proof sliding barrel along the elastic axis vertical layout, the flame-proof tube size about Ø8mm×30mm[4].In a safe state, the detonation isolation is carried out through the dislocation relationship among the electric detonator, detonator and booster explosive.

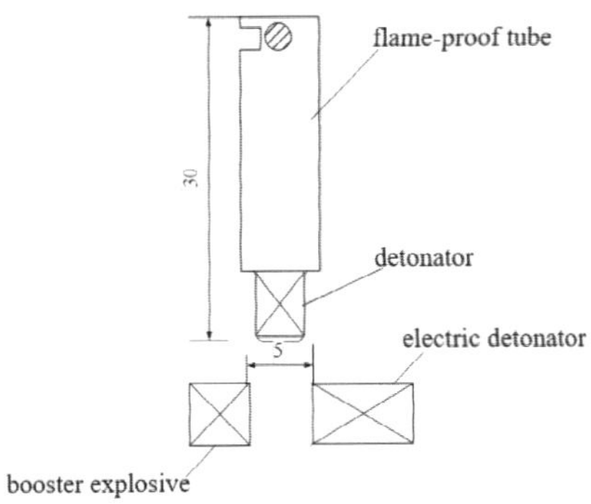

Figure 1. Schematic diagram of small penetration submunition fuze cartridge type flameproof device.

2.2. Properties of tungsten- copper alloys

Tungsten-copper alloy (W-Cu alloy) material has high strength, high density and low thermal expansion coefficient[5], widely used in aviation, aerospace, weapons and other military industries[6]. The higher the density, lower the expansion coefficient, and higher the tensile strength of the metal materials used in fuses flameproof plates, the better the flameproof performance[7]. These properties make W-Cu alloy have good characteristics of absorbing detonation, which is suitable for small size fuse deflector.

3. Ultra-thin fuse flameproof plate arranged vertically

In order to improve the firing rate of the new type of small penetration submunition, the double-path explosive isolation mechanism should be designed to increase the double-path redundant transmission sequence. The vertical layout of the detonation isolation mechanism is required because of the large overload of the target base. At the same time new small penetration submunition fuze explosion mechanism is to meet the requirements of flame-proof and to meet the requirement of high penetration overload and, in this paper, the vertical arrangement of fuze ultra-thin flame-proof board proposed composite structure with three layers of different metal materials (Figure 2), flame-proof institutions within the very narrow space using symmetric dual vertical arrangement, to meet the demand of the two-way combustion and flame-proof (Figure 3).

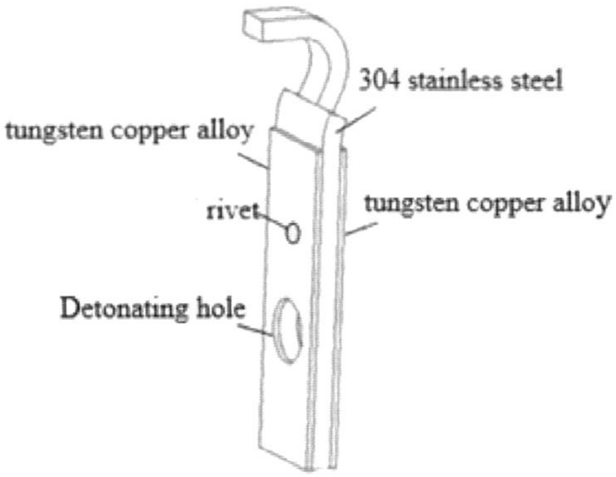

Figure 2. Ultra-thin fuse deflector arranged vertically.

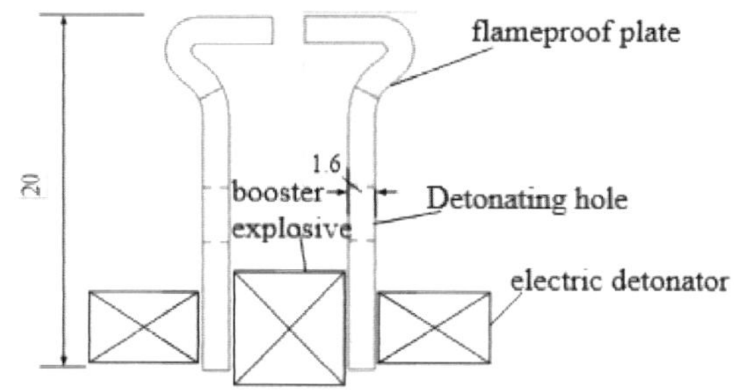

Figure 3. Schematic diagram of double channel layout of fuze ultra-thin type flameproof plate arranged vertically.

Its single overall size is 1.6mm thick, the vertical length is about 20mm, the upper and lower two layers are respectively 0.3mm W-Cu alloy plate, the middle layer is 304 stainless steel, between the three layers by riveting connection. In order to resist high penetration overload in the process of penetration, the bending Angle structure at the upper end of the middle layer was designed. The whole design of the flameproof plate has a detonation guide hole, and the electric detonator transmits the detonation charge through the detonation guide hole. Compared with Figure 1, the vertical placement of fuze ultra-thin type flameproof plate can reduce the detonation sequence of the first-order detonator, and replace it with the detonation hole, the structure can be more than doubled.

4. Simulation and test verification

4.1. Simulation verifies the capability of the ultra-thin vertical fuze flameproof plate against high penetration overload

Due to the small penetration submunition in the process of penetration by as much as 120,000 g of penetration overload, so the vertical arrangement of fuze ultra-thin flame-proof plate by simulating the resistance to penetration overload capacity, the overload curve as shown in Figure 4 [8]. If the maximum stress of the flameproof plate is less than the yield limit of the materials used (Table 1), the flameproof plate meets the requirements of high penetration overload. The finite element model is shown in Figure 5. The simulation unit system is mm-ms-g-MPa.

The material parameters used in the simulation are shown in Table 1:

Table 1. Material parameters selected for simulation.

Diaphragm material	Density (g/mm3)	Elastic modulus (MPa)	Poisson's ratio	Tensile strength (MPa)	Yield limit (MPa)	Shear modulus (MPa)
304 stainless steel	7.93×10^{-6}	194×10^{3}	0.3	520	205	58×10^{3}
Tungsten copper alloy	19.3×10^{-6}	410×10^{3}	0.33	1200	2000	136×10^{3}

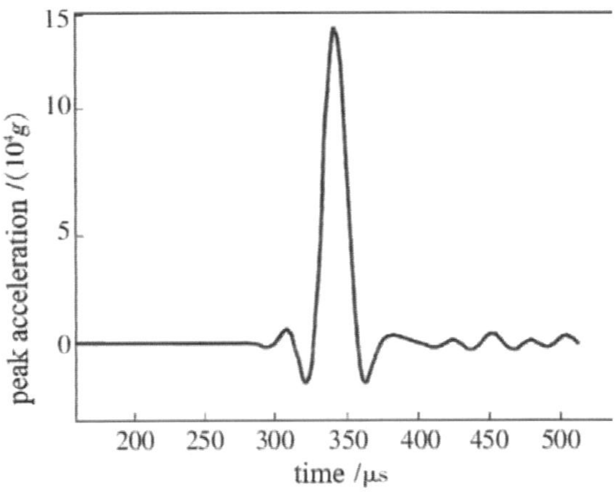

Figure 4. Penetration overload curve.

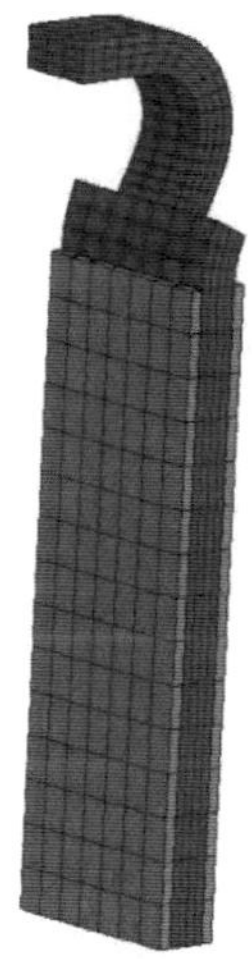

Figure 5. A simplified finite element model for calculation.

Under the condition of maximum penetration overload of 120,000 g, the forces of the vertically arranged fuze ultra-thin explosion-proof plate were simulated respectively. The simulation results are shown in Figure 6. The maximum stress appears at the inner bend of the upper end of the hook structure in the middle layer of the flameproof plate, which is 17.6 MPa, less than the yield limit of 304 stainless steel used in the middle layer, which is 205 MPa, and the structure will not fail.

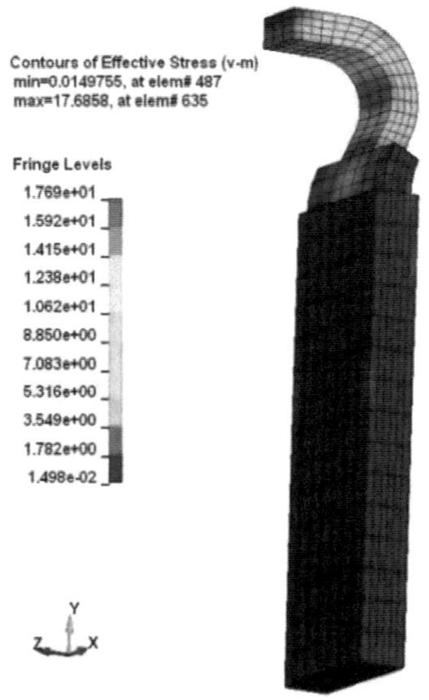

Figure 6. 120, 000 g stress cloud diagram of the deflector under overload penetration.

Maximum stress throughout the ultra-thin flame-proof plate are less than the yield limit of material used, so the vertical arrangement of fuze ultra-thin flame-proof plate can meet the small penetration ammunition drug resistance of 120,000 g penetration overload work requirements.

4.2. Simulation verifies the safety of ultra-thin fuses arranged vertically when 12m falling

In the service process, because the fall is likely to lead to the vertical arrangement of fuze ultra-thin flame-proof board security failure occurs, so drop of simulation verification, if the flame-proof board of maximum stress is less than the yield limit of material used (Table 1), the flame-proof plate meet the security requirement 12m falls. In the most dangerous case, the sag overload can be simulated by a curve whose peak value is 12,000 g and the corresponding time is 230μ s. The curve of sag overload is shown in Figure 7.

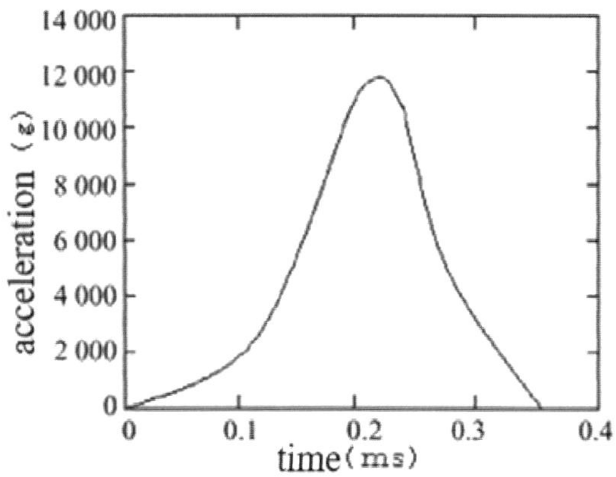

Figure 7. The overload curve of 12m falling.

The simulation results as shown in Figure 8, the maximum stress in the middle tier of the flame-proof institutions within the top hook structure of bend, 2.1 MPa, the material yield limit of 304 stainless steel is far less than the middle layer 205 MPa, so fall effect of overload of flame-proof board is minimal.

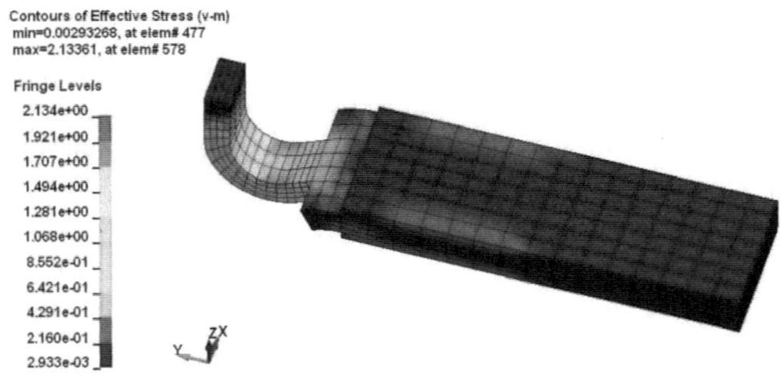

Figure 8. Stress clouds under 12m falling.

Throughout the ultra-thin flame-proof board maximum stress is less than the yield limit of material used, so the vertical arrangement of fuze ultra-thin flame-proof board hook upper middle tier structure in the bend of the safety of the 12m fall in dealing with the satisfied service.

4.3. The test verifies the explosion isolation safety of the ultra-thin fuses arranged vertically

Vertical arrangement of fuze ultra-thin flame-proof board's primary function is to ensure that the detonator in occasional case the advanced ignition, will not set off at the next lower level booster column, so as to avoid bullets medicine early burst and cause damage to our personnel and equipment. Therefore, the explosion isolation performance of the explosion isolation mechanism is tested and verified.

Test is new small drug penetration ammunition Ø 2.5mm size distribution by electric detonator, the electric detonator can effectively pierced the 3mm steel plate. In electric detonator with high overload resistance of ultra-thin fuze flame-proof $\ngtr$0.1mm agency clearance under the conditions of explosion-proof test, if the third layer of tungsten copper alloy plate is in good condition, the isolation plate of flame-proof performance meet the safety requirements.0.1mm

After the detonation of the electric detonator, the explosion isolation mechanism is disassembled layer by layer. The first layer of W-Cu alloy sheet, the middle 304 stainless steel sheet and the third layer of tungsten-copper alloy sheet are respectively shown in Figure 9 from left to right. It can be seen that the first layer of W-Cu alloy sheet is seriously damaged, and the middle 304 stainless steel layer has a slight dent caused by detonation of the detonator. The third layer of tungsten-copper alloy sheet is intact without any trace.

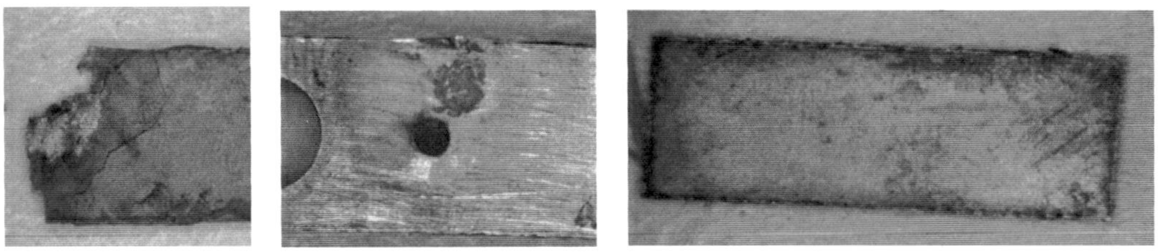

(a) **(b)** **(c)**

Figure 9. Details of each layer of the flameproof plate after the test.
(a) The upper layer plate; **(b)** The inter layer plate; **(c)** The under layer plate.

Test results show that although the high overload resistance of ultra-thin fuze explosion mechanism of the first layer is fragile, but absorption detonation energy effect is good, can effectively reduce the next layer of 304 stainless steel plate have been hit. Therefore, the whole vertical layout of the fuze ultra - thin type flameproof plate has good performance.

Comprehensive above the results of simulation and experiment results, which can be concluded that the overall size is about 1.6mm×20mm vertical arrangement of fuze ultra-thin flame-proof board, can resist 120,000 g penetration overload, 12m falling at the same time meet the explosion-proof security and safety.

5 Conclusion

In this paper, the vertical placement of the fuze ultra-thin type flameproof plate, the overall appearance of which is three layers of different metal materials vertical composite sheet structure, its overall size is 1.6mm thickness, vertical length about 20mm. The middle layer is made of 304 stainless steel, and the bend at the upper end of the middle layer is designed to resist penetration overload. The upper and lower layers are made of 0.3mm thick W-Cu alloy plates to absorb detonation energy of detonators. Verified by simulation, the vertical arrangement of fuze ultra-thin flame-proof plate can resist 120,000 g penetration overload, and satisfy the security of the 12m falls, and experiments verify its meet new small drug penetration ammunition flame-proof required by the security. Direction of further research for the vertical arrangement of fuze ultra-thin flame-proof plate can be applied to other ammunition fuze, at the same time, with the continuous development of material technology, the future may be a new kind of material, adopting the new materials of the vertical arrangement of fuze ultra-thin flame-proof plate to the structure of the smaller size and better flame-proof security.

6.Reference

[1] LI Dongjie, LI Huamei, TAN Liang and MU Dong. Overview on Fuze Electronic Safety and Arming Device Development [J]. *Journal of Detection & Control*,2015,4(37):1-6

[2] John Robbins. US NAVY OVERVIEW [C]//*49th Annual Fuze conference*. USA: NDIA, 2005

[3] Max Perrin. New Ssfety Requirements For Munitions Fuzing System Solutions [C]//*54th Annual Fuze Conference*. USA: NDIA, 2010

[4] Li Xiaofeng, Wang Yabin and Wu Bi. *Fuze of Penetration Ammunition*[M]. Beijing:National Defense Industry Press, 2016, first edition

[5] Liu Xiaoling. *Testing And Analysis Of Macro-Micro Mechanical Properties Of Tungsten Alloy*[D]. Chengdu: South Jiaotong University, 2003, 12

[6] Rong Jili, Yu Xinjian, Liu Bin and Hu Gengkai. FEM Simulation and Analysis of Tungsten Heavy Alloy Fragile Projectile Under Impact Dynamics [J]. *Transactions Of Beijing Institute Of Technology,* 2004, 3(24): 193-196

[7] *Fuze design manual* Editor. *Fuze design manual*[M]. Beijing：National Defense Industry Press, 1978, first edition

[8] Jin Hong, Tang Bo, Jin Shuyun and Chen Changxin, Wang Huan. Study on numerical simulation of characteristic parameters of penetration overload curves [J]. *China Measurement & Test*, 2015, 6(41): 125-128

IOP Conf. Series: Journal of Physics: Conf. Series **1209** (2019) 012004 doi:10.1088/1742-6596/1209/1/012004

Parametric design of MEMS fuze safety release insurance mechanism

F Q Zheng[1,4], W Z Lou[1], F F Wang[2] and D K Wang[3]

[1]National Key Laboratory of Electro-Mechanics Engineering and Control, School of Electro-Mechanics Engineering, Beijing Institute of Technology, Beijing 100081, China.
[2]Technology and Engineering Center for space Utilization, Chinese Academy of Sciences, Beijing 100094, China.
[3]Beijing Institute 0f Electronic System Engineering, Beijing 100854, China.

[4]Corresponding Author: Fuquan Zheng; email: zhengfuquan1994@163.com; Tel.: 18810715855

Abstract. At present, MEMS safety systems have the disadvantages of complex structure and easy to be affected by the electromagnetic environment of the battlefield. Based on this, this paper proposes a planar structure to realize remote protection of MEMS security systems. First, a theoretical analysis of the MEMS remote release insurance mechanism is carried out. It mainly calculates the centrifugal force of the centrifugal slider, the magnitude of the deflection of the threshold elastic beam caused by the active tooth collision fixed tooth, and the time when the active tooth of the safety mechanism is released by the fixed tooth. Then, the motion characteristics of the mechanical MEMS remote release insurance mechanism are analyzed by simulation. Through theoretical analysis and simulation analysis, the theoretical calculation results of the maximum deformation of the active tooth are in good agreement with the simulation results, and the error is less than 9.7%. The delay time can reach 1.33ms or more, which meets the requirements.

1.Introduction

The long-distance disarming function of the traditional safety system is generally implemented by means of gunpowder extension, clockwork, electric firework delay, ball rotor, belt, fusible alloy and quasi-fluid [1-2]. Due to the limitation of the processing technology of the MEMS safety system, it is required that the long-distance release insurance mechanism has the characteristics of a planar structure, a small size, and is easy to process and assemble. At present, MEMS safety systems often use a flat-type timepiece to remotely release the insurance mechanism or use a micro-sensor to detect the distance of the projectile from the muzzle, combined with the micro-actuator and form a long-distance disengagement mechanism with a key module structure on the MEMS safety system [3-5].

The long-distance release of the flat-type timepiece is more complicated and takes up a lot of space. The rotary pair connected with the escape wheel is a difficult point of processing [6]; An electronic remote disengagement mechanism is formed using a microsensor combined with a micro actuator and combined with a key module on the MEMS safety system. As shown in Figure 1 [7-8]. The structure is simple, the design form is diverse, the control precision is high, and the principle of high-density integration is more suitable. The disadvantage is that it is susceptible to the harsh electromagnetic environment of the battlefield. According to the process characteristics of MEMS and the characteristics of long-distance disarming mechanism, this paper proposes a planar structure to realize long-distance disarming of MEMS safety system.

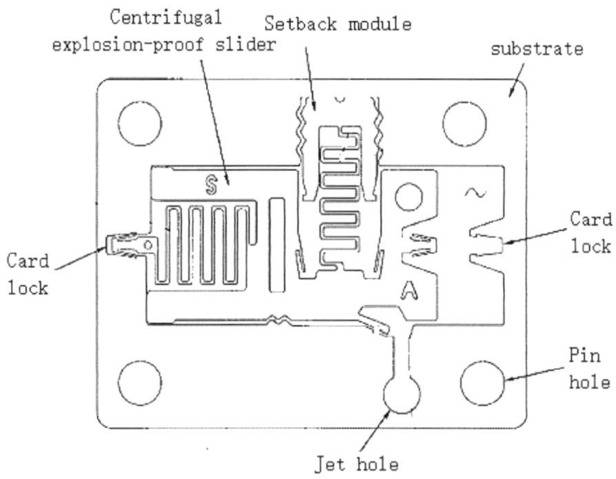

Figure 1. MEMS security system electric remote release insurance mechanism diagram.

2.Theoretical Analysis of Mechanical MEMS Remote Discharge Insurance Mechanism

2.1 Conception and parametric design of mechanical MEMS remote release insurance mechanism

Long distance insurance is a basic requirement for the design of MEMS safety systems. In this paper, a mechanical MEMS remote release mechanism is designed based on the silicon-based DRIE process, as shown in Figure 2.

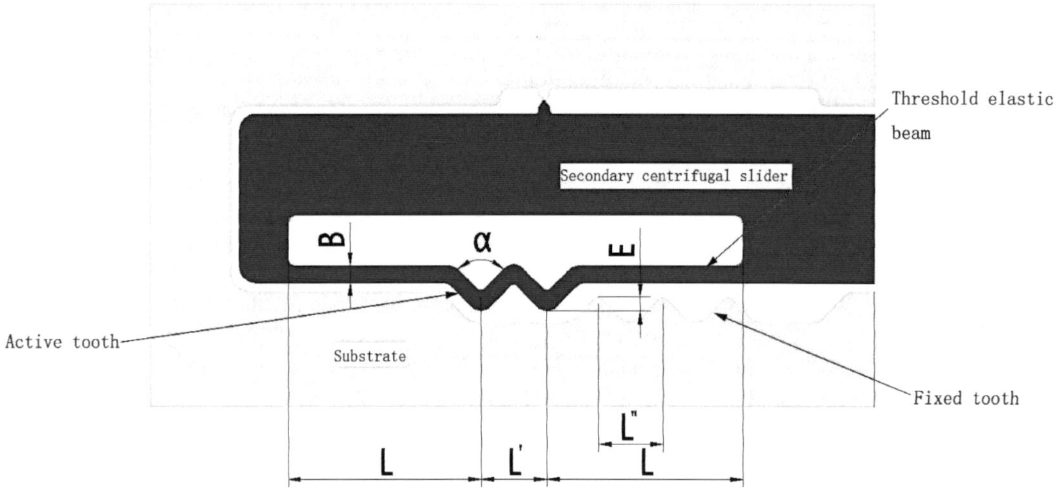

Figure 2. Mechanical MEMS remote release insurance mechanism diagram.

The elastic beam delay mechanism mainly includes a threshold elastic beam on the auxiliary centrifugal slider, the active tooth and the fixed tooth on the substrate, as shown in Figure 2. The action process is as follows: under the action of the centrifugal environment, the secondary centrifugal slider is moved by the centrifugal force. The active teeth on the secondary centrifugal slider are interlaced with the fixed teeth on the substrate, and the movement of the secondary centrifugal slider is blocked. Since the threshold elastic beam and the corresponding cavity are designed on the auxiliary centrifugal slider, the active teeth on the auxiliary centrifugal slider move passively toward the cavity, thereby moving the auxiliary centrifugal slider to the outside. When entering the next tooth, the active tooth rebounds to its original position by the influence of the threshold elastic beam. The active tooth reciprocates up and down, and the secondary centrifugal slider moves slowly to the outside to achieve the purpose of delaying for a certain time. Its purpose is a planar design to achieve a short delay function of the MEMS security system. Parametric design for the main variables.

Table 1. Parameterization of mechanical MEMS remote disarming mechanism.

num	1	2	3	4	5
Line width B/mm	0.1	0.08	0.1	0.1	0.1
inclination α /°	90	90	70	90	90
Arm length L/mm	1.2	1.2	1.2	2.4	1.4
Gap E/mm	0.03	0.03	0.02	0.03	0.03
Number of active teeth	2	2	2	2	1
Active tooth spacing L'/mm			0.4		-
Fixed tooth spacing L''/mm			0.4		

2.2 Theoretical calculation of mechanical MEMS remote release insurance mechanism

For the remote release of the insurance mechanism, the theoretical calculations are carried out on two aspects: (1) the centrifugal slider is subjected to centrifugal force, which causes the active tooth to collide with the fixed tooth to cause the deflection of the threshold elastic beam; (2) the long-distance release mechanism of the active tooth through the fixed tooth.

2.2.1 Centrifugal slider deflection calculation. Assume that the model shown in Figure 2 assumes that the model does not move up and down and affects the motion performance of the elastic beam. Establish a force analysis diagram of the mechanical MEMS remote release insurance mechanism, as shown in Figure 3.

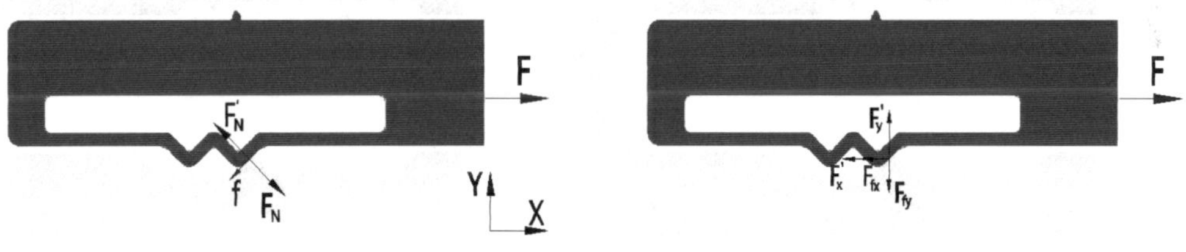

Figure 3. Force diagram of mechanical MEMS remote release mechanism.

As shown in Fig. 3(a), the main force is the centrifugal force F generated by the rotation of the centrifugal slider by the projectile. The centrifugal slider is in motion, and the active tooth is in contact with the passive tooth. The active tooth gives the passive tooth ramp a pressure perpendicular to the passive tooth $F_N(w)$. $F_N(w)$ is the force that changes with the deflection of the elastic beam delay mechanism, while the active teeth are subjected to reaction forces F_N'. Where $F_N' = F_N(w)$ is the friction generated by the relative motion of the active and passive teeth. The centrifugal force generated by the projectile rotating centrifugal slider is as follows:

$$F = mr\omega^2 = mr(\frac{2\pi n}{60})^2 \tag{1}$$

Where m is the mass of the centrifugal slider, r is the eccentricity of the centrifugal slider, and n is the rotational speed of the projectile.

Applying the engineering mechanics cantilever beam deflection curve equation, the mechanical MEMS remote release insurance mechanism is analyzed, as shown in Figure 4:

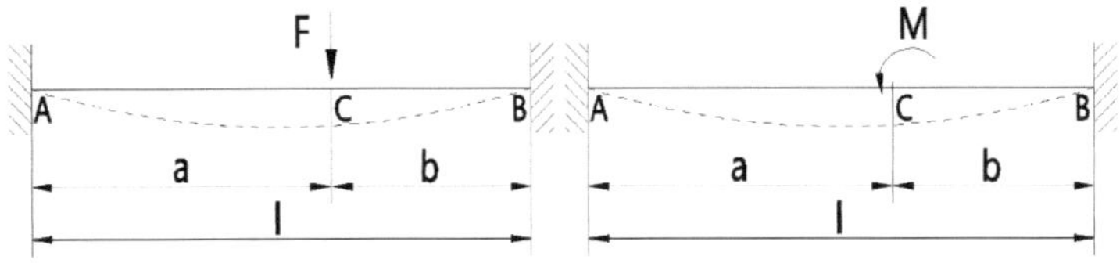

Figure 4. Schematic diagram of deflection analysis of elastic beam.

Can get the maximum deflection value at node C:

$$w_{SC\max} = w_{SC1\max} + w_{SC2\max} \tag{2}$$

In the formula:

$$\begin{cases} w_{SC1\max} = \dfrac{4F\left(L - L'/4\right)^3\left(L + 5L'/4\right)^3\left(\sin(\alpha/2) - \mu\cos(\alpha/2)\right)}{EHB^3(2L+L')^3\left(\cos(\alpha/2) + \mu\sin(\alpha/2)\right)} \\ w_{SC2\max} = \dfrac{2}{EHB^3(2L+L')}\left(-3M_A - R_A\left(L + 5L'/4\right)^3\right) \end{cases} \tag{3}$$

Calculated by calculation $w_{SC\max} >> w_{SC2\max}$,which is $w_{SC\max} = w_{SC1\max}$. According to the geometric relationship shown in Figure 4, the displacement of the slider motion is obtained.

$$S_S = w_{SC\max}\tan(\alpha/2) \tag{4}$$

In the case of F_x=0, the centrifugal slider has a speed and the speed is the maximum. After that, the speed gradually decreases until the speed drops to zero, and the deflection reaches a maximum.

Assuming that the motion of the active tooth contact passive tooth is from uniform acceleration to uniform deceleration, the average acceleration of the process of uniform acceleration is $\bar{a}_A = a_{A\max}/2$ =F/2m, and the average acceleration of the process of uniform deceleration is $\bar{a}_D = a_{D\max}/2$.

$$S_D = w_{MC\max}\tan(\alpha/2) - S_A = \left(w_{MC\max} - w_{SC\max}\right)\tan(\alpha/2) \tag{5}$$

Calculated as follows:

$$w_{MC\max} = -\frac{c_2}{c_1} = \frac{8F\left(L - L'/4\right)^3\left(L + 5L'/4\right)^3\left(\sin(\alpha/2) - \mu\cos(\alpha/2)\right)}{EHB^3(2L+L')^3\left(\cos(\alpha/2) + \mu\sin(\alpha/2)\right)} \tag{6}$$

2.2.2 Exercise time estimation. In this design, the delay time is mainly achieved by the active tooth collision passive tooth reducing the motion speed.The relationship between displacement and velocity, acceleration, and time is as follows:

$$S = v_0 t + \frac{1}{2}at^2 \tag{7}$$

(1)Active tooth first contact with passive teeth

According to the previous analysis of the deflection of the dynamics, it is known that from the first contact of the active tooth to the passive tooth, to the escape of the passive tooth, an acceleration and a deceleration process are experienced. It is assumed to be a uniform acceleration process and a uniform deceleration process. We can get:

$$t_{1A}=\sqrt{\frac{2E\tan\left(\frac{\alpha}{2}\right)}{r_0\left(\frac{2\pi n}{60}\right)^2}} \tag{8}$$

Where r_0 is the initial eccentricity of the centrifugal slider.

It can be seen from the analysis of the kinetic mid-deceleration process that the uniform deceleration process and the uniform acceleration process calculated by the above formula are inverse processes. Therefore $t_{1S}=t_{1A}$, i.e. $t_1=2t_{1A}$.

(2) active teeth second contact with passive teeth

After the first contact of the active tooth and the acceleration of the intermediate portion, the second contact with the passive tooth has an initial velocity. Therefore, the initial velocity at the second contact is first requested, and then the time is calculated by the formula (7).

It is assumed that the acceleration of the middle portion of the first contact and the second contact takes the average value of the maximum value and the minimum value. The velocity is 0 when the first contact is passed and the passive tooth passes. The middle part of the exercise time is

$$t_1'=\frac{v_2-v_1}{\bar{a}_1'} \tag{9}$$

Where $\bar{a}_1'$ is the average acceleration of the middle part and v_2 is the initial speed of the second contact passive tooth.

When the active tooth touches the passive tooth for the second time, although the two teeth of the active tooth are in contact with the passive tooth at the same time, since the two teeth of the active tooth are connected, only the force of one tooth can be calculated. After analysis, t_2 satisfies the following relationship.

$$E\cdot\tan\left(\frac{\alpha}{2}\right)=v_0t_2+\frac{1}{2}\bar{a}_2t_2^2 \tag{10}$$

$\bar{a}_2$ is the average acceleration of the active tooth in the second contact with the passive tooth process. The total time the safety slider has moved into place is:

$$\mathrm{t}=t_1+t_1'+t_2 \tag{11}$$

3. Simulation analysis of mechanical MEMS remote release insurance mechanism

Through the above analysis, theoretical calculation and parametric design of the mechanical MEMS remote release insurance mechanism and optimization mechanism are carried out. This section mainly analyzes the dynamics of the working process of the mechanical MEMS remote release insurance mechanism, and analyzes the motion characteristics of the mechanical MEMS remote release insurance mechanism through simulation.

In the dynamic simulation of mechanical MEMS remote release insurance mechanism, the Pro/E software is used to establish the three-dimensional geometric model of the mechanical MEMS remote release insurance mechanism and the optimization mechanism, and the finite element model of the model is established by HyperMesh software. As is shown in Figure 5.

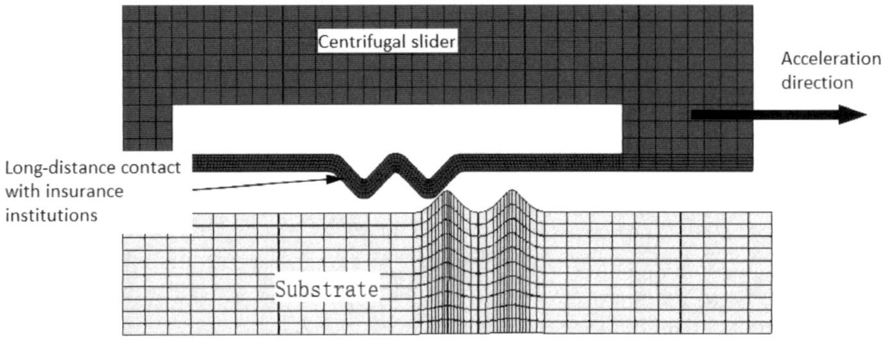

Figure 5. Finite element model of mechanical MEMS remote release insurance mechanism.

In this mechanism, the centrifugal slider and the substrate are made of silicon material. The material parameters in the model are listed in Table 2:

Table2. Material mechanics properties used in simulation.

Material name	Density ρ kg/m^3	Elastic Modulus E GPa	Poisson ratio ν
Si	2.3×10^3	180	0.3

3.1 Deflection simulation

Combining the parametric model listed in Table 2 with the structure shown in Figure 2, a finite element model as shown in Figure 5 is created and a constant size of 30 000 g is applied to the centrifugal slider. Adjust the gap E between the active tooth and the passive tooth in Figure 2, and explore the displacement of the active tooth in the working condition to overcome the displacement of the passive tooth. Combined with calculation and simulation, it can be obtained as shown in Table 3. In the theoretical and simulation cases, different parametric models produce maximum deflection values and comparison tables.

Table 3. Maximum deflection value comparison.

Comparison project	1	2	3	4	5
quality (m/kg)	2.88×10^{-6}	2.836×10^{-6}	2.866×10^{-6}	2.84×10^{-6}	2.868×10^{-6}
Theoretical maximum deflection (mm)	2.86×10^{-2}	5.50×10^{-2}	2.0×10^{-2}	0.20	3.23×10^{-2}
Simulation maximum deflection (mm)	2.92×10^{-2}	5.64×10^{-2}	2.1×10^{-2}	0.213	3.37×10^{-2}
error	2.1%	2.5%	5.0%	6.5%	4.3%

It can be seen from Table 3 that the simulation results about the deflection are in good agreement with the theoretical calculation results. The error of each number of the original organization is less than 6.5%. The reason is that when the simulation, the active tooth and the passive tooth have a certain gap at the initial moment, and there is a velocity at the time of contact, so that the force of the active tooth is increased, and the simulated deflection is greater than the theoretical calculation deflection.

3.2 Exercise time simulation

To simplify the calculation process, a constant 30 000 g acceleration is applied to the centrifugal slide. The theoretical calculation results of the time required for the active tooth to pass through two passive teeth are shown in Table 4.

Table 4. Theoretical calculation results of the time required for the active tooth to pass through 2 passive teeth.

Time/μs	1	2	3	4	5
t_1	276	283	231	730	293
t_1'	497	480	503	365	495
t_2	19	38	13	183	22
t_2'	32	32	32	320	32
Total time/μs	824	833	779	1330	842

As can be seen from Table 4, the most important factor affecting the exercise time is the value of the gap E. Under the same conditions, the larger the gap E, the longer the motion is displaced. As the displacement of the active teeth increases, the required time passes through the same distance, and the time required is shorter and faster.

Combining the parametric model listed in Table 1 with the structure shown in Figure 5, the active tooth gap is taken as the simulated value shown in Table 3 The simulation results of the displacement-time and speed-time of the active tooth passing through two passive teeth are shown in Figure 6.

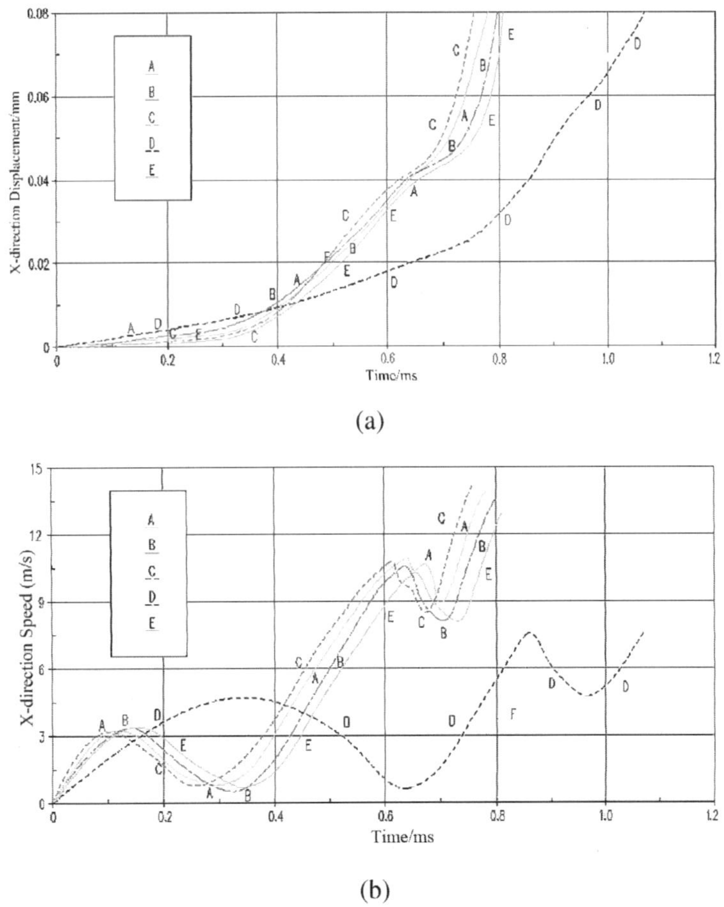

Figure 6. Simulation results of different structural parameters of the active tooth through two passive teeth.

According to the displacement-time and velocity-time simulation results shown in Figure 6, the simulation results are in good agreement with the theoretical calculation trends shown in Table 4 Overall, after the same displacement, the simulation time is shorter than the theoretical calculation. According to Figure 6(b), after the active tooth collides with the passive tooth, the theoretically calculated velocity is 0m/s, and the speed calculated by the simulation is obtained. More than 0m/s, due to the existence of the storage speed, the average speed of the simulation results is larger than the theoretical calculation, and the time for the same displacement is shorter. The active tooth has a maximum delay time of 1.33 ms through 2 passive teeth.

4.Conclusions

The parametric design of the MEMS fuze safety insurance mechanism is carried out. Through theoretical analysis and simulation analysis, the maximum deformation amount and the delay time parameterized design law are studied. The theoretical calculation results of the maximum deformation prediction are in good agreement with the simulation results. The error is less than 9.7%; the fourth scheme has the best effect, and its parameters are line width 0.1mm, inclination angle 90°, tooth arm length 2.4mm, gap 0.03mm, number of active teeth 2, active tooth spacing 0.4mm, fixed The tooth pitch of 0.4mm can make the delay time reach 1.33ms or more, which meets the requirements.

Acknowledgments

The research is supported by the State Key Laboratory of Mechatronics Engineering Control and sponsored by National Project (20160229150)

5.References

[1] F Wang, W Lou, Y Xiong, F Liu, D Wang, and X Jin. 2015. The parametric analysis of the Lock-Releasing mechanism in MEMS safety and arming device. *IEEE International Conference on Nano/micro Engineered & Molecular Systems*. 645-646, 986-989.

[2] X Jin, W Lou, F Wang, D Wang, M Guo and X Dai. 2015. The MEMS experiment and numerical simulations based on the explosion protection of micro-scale vehicles model. *IEEE International Conference on Nano/micro Engineered & Molecular Systems.*

[3] Tu H M, Sun Z L, Qian Y Pand Q Liu. 2017. Reliability analysis and improvement of mems-based safety and arming device in fuze. *Acta Armamentarii. 38*(4): 664-672.

[4] Ma B H 1984. Fuze Structure and Function principle. *National Defense Industry Press.*

[5] X Li, Y Zhao, and T Hu.2015. Design of a large displacement thermal actuator with a cascaded V-beam amplification for MEMS safety-and-arming devices. *Microsyst. Technol.* 21: 2367–2374.

[6] F Wang, W Lou, F Liu, D Wang, X Jin and J Lu. 2015. The calculation method of reliability based on MEMS Safety and Arming device. *IEEE International Conference on Nano/micro Engineered & Molecular Systems.*

[7] Shaeffer, D K 2013. MEMS inertial sensors. *A tutorial overview. IEEE Commun. Mag.*51: 100–109.

[8] Allameh, S.M. 2003. An introduction to mechanical-properties-related issues in MEMS structures. *J. Mater. Sci.*38: 4115–4123.

IOP Conf. Series: Journal of Physics: Conf. Series **1209** (2019) 012005 doi:10.1088/1742-6596/1209/1/012005

Effect of annealing process on microstructure and electrical properties of ni films

D R Wang[1], Z Huang [2], L Wang [2] and Z l Jiao [2]

[1,2] Nanjing Electronic Devices Institute, No.524, zhongshan road,Nanjing, Jiangsu Province, China

455549612@qq.com

Abstract. By DC magnetron sputtering,the Ni films of 600 nm thickness were successfully deposited on BF33 glass substrates .Then annealing at different temperature from 150 to 450°C and different holding time from 3 to 6 hours under vacuum conditions using bonding machine.The surface morphology and structure of Ni films were analyzed by scanning electron microscopy(SEM), films surface roughness and resistivity were investigated by atomic force microscope(AFM) and four-probe tester respectively. The results show that the annealing process contributes to the refinement of the grains.And the films roughness reaches the best level at the annealing temperature of 300 ° C and the roughness data with holding time of 6h is 0.678 nm. Above all ,As the annealing temperature increases, the films resistivity decreases significantly .The surface roughness decreases and the resistivity gradually decreases.

1. Introduction

Ni films resistor is a new type of temperature measuring element researched in recent years. It has large temperature coefficient of resistance, low melting point, easy film formation and low price. It has been widely used in various temperature sensors and thermal flow sensors. In addition, the nano-Ni film can be deposited by using a flexible polymer material as a substrate to prepare an ideal polymer materials for developing electromagnetic wave shielding, fiber sensors, solar panel, and so on[1-2].

The film prepared by magnetron sputtering has uniform film structure, compactness and good performance, and the film adheres firmly to the substrate, and has obvious advantages in applications such as conduction, antistatic, radiation protection and electromagnetic shielding. Since the surface morphology of the metal film is very important in the fabrication of some film resistors, it is generally believed that when the thickness of the film can be compared with the mean free path of the electron, the effect of the surface morphology on the resistivity of the metal films has a size effect, but when the film is very thick, the effect is not obvious[3-5]. The surface morphology of the film directly affects its performance. In addition to the macroscopic properties of film thickness and roughness, the microscopic morphology of the film, such as grain size, voids and cracks, and film structure, can better reflect the quality of the film. The effects of annealing temperature and annealing time on the surface morphology, film resistivity and surface roughness of sputtered Ni films were investigated.

In this paper, the Ni films was prepared by DC magnetron sputtering.The film sample was annealed in a vacuum environment of the bonding machine. Different annealing temperatures and holding times were selected for the annealing process. The surface morphology, film resistivity and surface roughness of the Ni films were tested by scanning electron microscope(SEM), atomic force microscope(ATM) and four-probe tester Then the optimum annealing conditions were obtained.

2. Experiment Procedure

2.1. Sample preparation

In this experiment, the Ni films was prepared by magnetron sputtering. In order to reduce the thermal stress caused by the mismatch of thermal expansion coefficient between the substrate and the film, a BF33 glass with a thermal expansion coefficient close to the substrate material was selected, and a high purity nickel target was used(99.99%). The distance between the target and substrate is 63mm, the sputtering gas is high purity argon (argon purity is 5N), the sputtering pressure is $3*10^{-3}$mbar.After wet cleaning process,the substrate should be kept horizontal. Before sputtering, a Pre-sputtering process for 2 min should be done to remove contaminants on the surface of the target, a negative bias is used to enhance the bonding strength between the film and the substrate, and the thickness of the sputtered Ni films is controlled to be about 600 nm.

2.2. Annealing experiment

The prepared films samples were subjected to in a vacuum environment using the bonding machine with the vacuum of 10^{-5} mbar. The experiment is divided into multiple groups. The first group of annealing temperature is set to 150, 200, 250, 300, 350, 400, 450°C. The second group is set to same annealing temperature, but with different holding annealling time of 3, 4 5, 6h respectively. The principle of the annealing treatment in this experiment is that the particles are re-obtained by thermal annealing to perform grain reconstruction. The grain size and spacing of the film directly affect the temperature coefficient of the film.The annealed experimental samples were tested by SEM,AFM and four-probe tester.

3. Results and Discussion

3.1. Study on Surface Morphology of Ni Films

The film samples before and after annealing were observed by SEM, and the surface morphology of Ni films at different annealing temperatures was taken. From Figure 1(a) we can be see before annealing that the Ni films consists of nearly regular block particles, the particle size and distribution are relatively uniform, and the gap is small, so this show that the Ni films belongs to a dense films. Figure 1(b)-(f) shows after different annealing temperature processing that the surface morphology of Ni films. It was found that the films was better than before annealing, and there were no defects such as cracks and voids. It proves that he slow annealing process contributes to the refinement of the grains, resulting in a reduction in the film gap and a more compact structure.Figure 1(f) is the morphology of Ni films in the case of rapid annealing at 450 °C, because rapid annealing accelerates the agglomeration of crystal grains,so this annealing condition resulting in larger grain size, poor compactness of the structure, and cracking in some film regions.

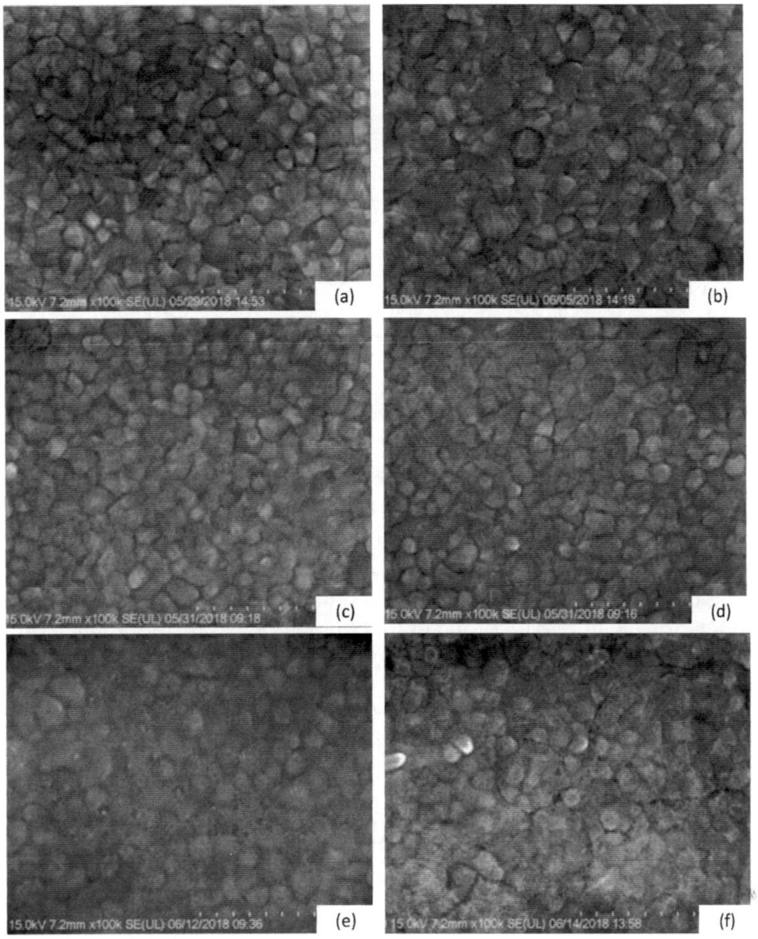

Figure 1.SEM microphotographs of the Ni films annealed at different temperature for 3h:(a) as deposited;(b) annealed at 150℃;(c) annealed at 200℃; (d) annealed at 250℃;(e)annealed at 300℃;(f) annealed at 450℃..

3.2. Study on Surface Roughness of Ni Films

3.2.1. Effect of annealing temperature on roughness. Figure 2 shows the roughness data at different annealing temperatures. The results show that as the annealing temperature increases, the surface roughness decreases first and then increases. The reason is that as the annealing temperature increases, the grain size increases, and after the atoms on the surface of the film gain energy, they migrate from the high potential point to the low potential point, and the defects such as film voids, dislocations, and vacancies are reduced. So the surface is smoother. At the annealing temperature of 300 ° C, the atoms get enough energy to fully migrate, at which point the roughness reaches the best level[6].

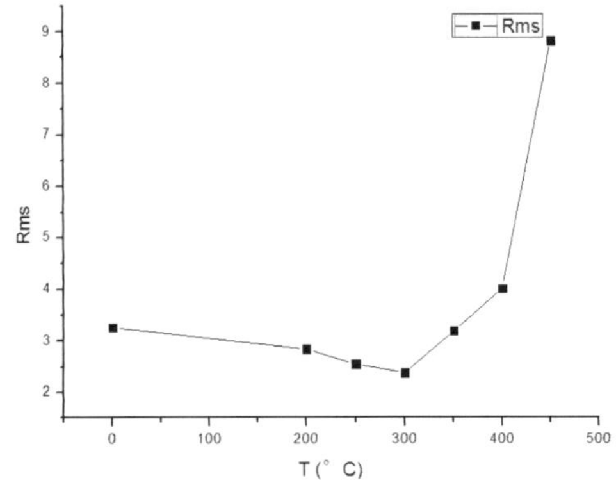

Figure 2. Roughness vs the temperature.

3.2.2. Effect of annealing time on roughness. Figure 3 shows the roughness data at different annealing holding time. From the figure we can see that the surface roughness continued to decrease with the increase of annealing time. The roughness data with holding time of 6h is 0.678 nm. Figure 4 are the AFM image of the Ni films taken before and after annealing. It can be seen that the zigzag surface of the magnetron sputtered Ni films has a high and low undulating state distribution, which is obviously superior to the undulation state before annealing. It indicating that the annealing process helps to improve the surface morphology, thereby reducing the surface roughness of the film to some extent.

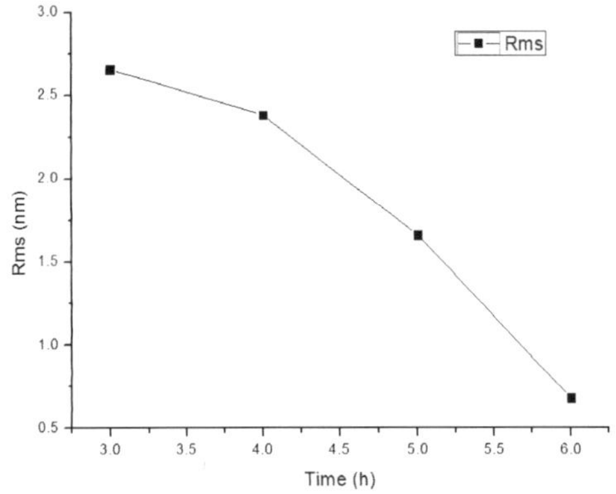

Figure 3. Roughness vs the time.

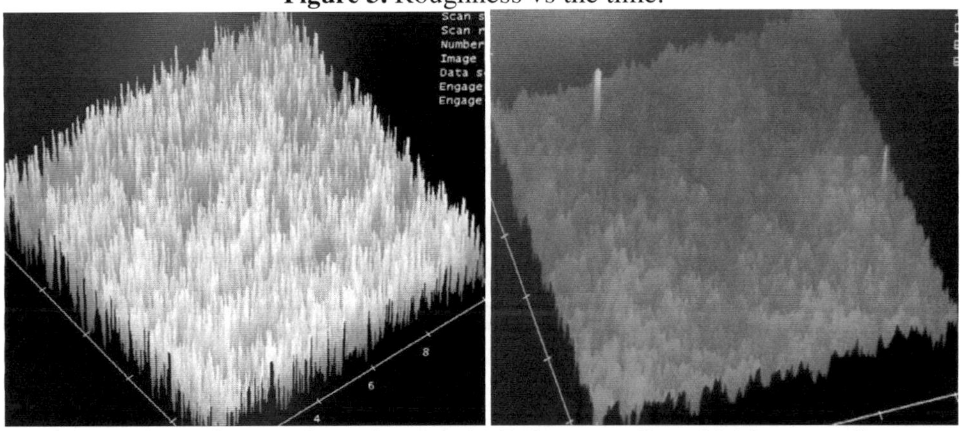

Figure 4. Surface roughness images: (a) Before annealing; (b)After annealing.

IOP Conf. Series: Journal of Physics: Conf. Series **1209** (2019) 012005 doi:10.1088/1742-6596/1209/1/012005

3.2.3. Effect of annealing on film resistivity. According to the electrical characteristics of the sheet resistance, it is generally composed of two parts[7-9]: one part is the resistance generated by the transmission of electrons between the crystal grains, and the other part is the electric resistance generated by the electrons transmitting inside the crystal grains. It is assumed that the inside of the grain is a continuous film transmission, and between the grains is activated tunneling.Resistivity between grains:

$$\rho_T = \frac{4}{3}\cdot\frac{r}{b^2}\cdot\frac{1}{e}\cdot\exp(\frac{E_s}{kT})\cdot e^{\frac{4\pi\cdot b}{\square}\sqrt{2m(W-E_k)}} \tag{1}$$

In formula (1), ρ_T is the resistivity between grains, E_s is the surface activation energy, k is the Boltzmann constant, T is the temperature, r is the grain radius, W is the tunnel barrier, and b is the grain. The distance between E_k is the electron kinetic energy and h is the Planck constant. Internal resistivity of the grain:

$$\rho_n = \frac{C\cdot T}{4\theta_D}\cdot\frac{4\cdot\lambda}{3d\cdot ln(\lambda/d)}(d \ll \lambda) \tag{2}$$

In the formula (2), ρ_n is the resistivity between the crystal grains, θ_D is the Debye temperature, C is a constant, T is the temperature, λ is the electron mean free path, and d is the film thickness.

It can be seen from the formulas (1) and (2) that the structure, size and spacing between the inside and the grains are different; the resistivity is also different. According to the electrical properties of the film resistor, the resistance increases with the increase of temperature due to the electron scattering inside the grain, and it has a positive temperature coefficient. The electrons are tunneled between the crystal grains, and the resistivity decreases with temperature and is negative coefficient. The final temperature coefficient of the sheet resistance is the result of competition between these two trends.

The resistance of the Ni films is an important parameter in the sensor design. In this paper, the sheet resistance Rs of the Ni films is measured by a Four-probe tester, and the resistivity of the sputtered Ni films is calculated according to the formula (3). Where ρ is the film resistivity and d is the film thickness, and the same film thickness is obtained by controlling the process parameters.

$$\rho = \mathrm{Rs}\cdot d \tag{3}$$

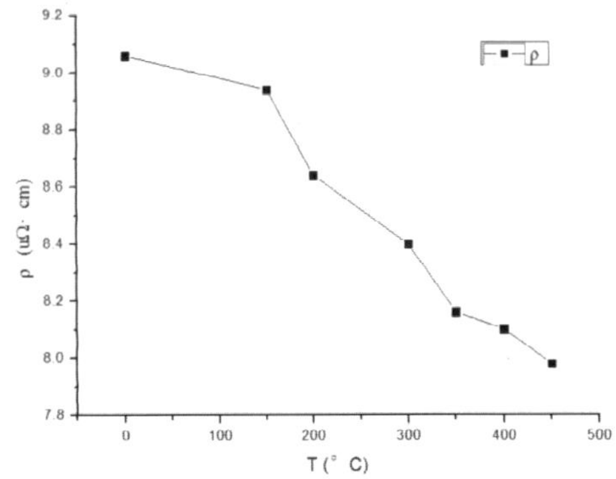

Figure 5. Film resistivity vs temperature.

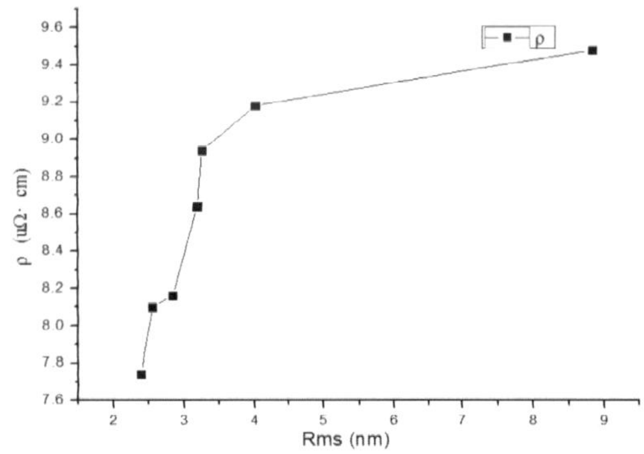

Figure 6. Film resistivity vs surface roughness.

The annealed samples was then tested. The annealing temperatures were set to 150, 200, 300, 350, 400, and 450 ° C respectively. Figure 5 shows the resistivity changes with the annealing temperature changes. From the figure it can seen that the annealing temperature and the resistivity are basically linear. As the annealing temperature increases, the resistivity of the Ni films decreases significantly. Because the metal films is annealed, the higher annealing temperature causes the fine particles in the film to merge into larger grains, the grain size increases, the intergranular area decreases, the degree of crystallization of the film increases, and the crystallization defect decreases. So the internal stress in the film is reduced, the grain boundary and surface scattering effects was weakened, the scattering effect of the carrier is weakened, the mobility is increased, the resistivity of the film is decreased, then the electrical conductivity is improved.

Figure 6 shows that effect between resistivity and surface roughness of Ni films. As shown, as the surface roughness increases, the resistivity of the film gradually increases. This is related to the growth structure of the Ni films. The magnetron sputtered Ni films grows in a columnar crystal structure, and the particle size is uneven in the direction perpendicular to the surface of the film. The particles of different heights in the longitudinal direction and different sizes in the lateral direction which result in different film surfaces roughness, which affects the transport of electrons in the film, resulting in changes in film resistivity.

4. Conclusions

The Ni films of 600nm was successfully prepared by magnetron sputtering. The effects of annealing temperature and annealing time on the surface morphology, film resistivity and surface roughness were investigated . The results show that the annealing process contributes to the grain refinement. The Ni films after annealing process is better than before annealing, and there are no defects such as cracks and voids. The grain size increases, the film gap decreases, and the structure is more compact. As the annealing time is lengthened, the surface roughness of the film continues to decrease. When the annealing temperature is 300 ° C, the atom obtains sufficient energy to fully migrate. At this time, the roughness data at 6 h of the holding time is 0.678 nm, and the roughness reaches the best level. In addition, the study found that the annealing temperature and resistivity are basically linear. As the annealing temperature increases, the resistivity of Ni films decreases significantly. As the surface roughness increases, the resistivity of the film gradually increases.

5. Acknowledgments

The authors would like to thank everybody of the research group help and support.

6. References

[1] RizzoA， Taglienate M A， et al .Structure and optical Properties of Silver Thin films Deposited By RF Magnetron Sputtering ,*J.Thin Solid Films*,2001,396(1):29-35

[2] Ya-Qi Hou， Da-Ming zhuang ,Gong Zhang ,Ming Zhao,Min-Sheng Wu, Influence of annealing temperature on the properties of titanium oxide thin film ,*J. Surface Science* 218 (2003), pp. 98-106

[3] Y.Huang, H.Qiu, L.Pan,et al. Effect of annealing on compositional , structure and electrical properties of Au layer grown on different thickness Cr layer, *J. Univ. Sci. Technol.* Beijing, 11(2004),No.3,p235

[4] Wang P H， Sun Z Q. Effect of annealing temperature on the microstructure and optical constants of Al film,*J.Journal of Anhui University (Nature Science Edition)*,2005,29(3),pp33-36

[5] Tang Wu,Deng Longjiang,Relationship between Resistivity of Metallic Film and Its Surface Roughness,Residual Stress,*J.Rare Meterials and engineering*, 2008, 37(4), 617 -620.

[6] Y.Huang, H.Qiu, L.Pan,et al.Effect of annealing on compositional ,structure and electrical properties of Au layer grown on different thickness Cr layer, *J.Univ.Sci. Technol.*Beijing, 11(2004), No.3,p235

[7] XIE Sheng,Hou Yu-wen,CHEN Chao,et al,Analysis the influence of Annealing on the resistance of NiCr thin film,*J.Journal of Huaqiao University*(Natural Science),2009,30(1),P27-29

[8] Wang Jing,YU huahua,Study on surface roughness and resisitivity of nano Ni-Al by magnetron sputtering,*J.Journal of Xi an polytechnic University*,2009,23(4),P150-153

[9] Xiao dingquan, zhu jianguo, et al. Thin film physics and devices , Beijing,2011,pp.140-164

Design, fabrication and testing of gain SPR sensor chip

L S Qin[1], X Chen [1], L L Zhang [2] and D F Cui [3]

[1]University of Chinese Academy of Sciences, State Key Laboratory of Transducer Technology, Institute of Electronics, Chinese Academy of Sciences, Beijing, CHINA

[2]University of Chinese Academy of Sciences, State Key Laboratory of Transducer Technology, Institute of Electronics, Chinese Academy of Sciences, Beijing, CHINA

[3]University of Chinese Academy of Sciences, State Key Laboratory of Transducer Technology, Institute of Electronics, Chinese Academy of Sciences, Beijing, CHINA

xchen@mail.ie.ac.cn

Abstract. The gain Surface plasmon resonance (SPR) chip can compensate for the SPR loss, which can improve the sensitivity of the SPR analysis system. In this paper, we design an SPR chip that includes a glass substrate, a gain layer, and a metal layer. Wherein the gain layer is a film obtained by adding IR-140 fuel to different concentrations of methylene chloride in which polymethyl methacrylate (PMMA) is dissolved. Then, we prepared a series of chips with a specific gravity of 0.4%, 0.8%, 1.2%, and 1.6% for the IR-140 in the gain dielectric layer. The fitting effect of the 2nd to 11th order polynomial fitting on the SPR curve is studied, and the SPR curve in this paper is processed by the preferred ninth order polynomial fitting. Finally, by studying the relationship between the thickness of the gain layer and the absorption peak generated by the SPR chip, a chip having an IR-140 dye specific gravity of 0.4% is preferred. The gain layer of the gain SPR chip is also expected to be further studied to further increase its sensitivity.

1. Introduction

SPR sensing technology is a physical optical phenomenon called Surface Plasmon Resonance (SPR). When a beam of light is directed from a refractive index to a light-diffusing medium having a refractive index, total reflection occurs when the incident angle is greater than a certain critical angle C. Due to the presence of a thin film of a suitable thickness on the SPR chip, the incident light illuminates the surface plasmon (SPW) along the surface of the metal conductor as it is totally reflected through the prism. Surface plasmon resonance (SPR) sensing technology is based on surface plasmons (SPPs).That has always been a hotspot in refractive index sensors due to its high resolution, high sensitivity, real-time and no markage. SPPs can be excited by an attenuated total reflection (ATR) based Otto or Kretschmann prism coupling structure [1-4]. However, the inherently high loss of SPPs limits the signal-to-noise ratio (SNR) and sensitivity of SPR sensors. The loss of SPPs can be compensated by using a gain medium in the vicinity of metals of different geometries, or by adding an active layer to the SPR sensor based on phase detection [5-7]. Recently, a five-layer SPR sensor model based on a gain dielectric layer has been proposed [8]. This sensor not only compensates for SPPs losses and can even amplify it. However, this chip has five layers, the structure is too complicated and requires an optical pump.

In this paper, a three-layer SPR sensor chip is proposed and fabricated and tested. The fabricated chip can improve the signal-to-noise ratio (SNR) of the SPR sensor and compensate for the loss of SPPs while finding that the sensor has a wider gradient measurement range [9–10]. It is expected that the ninth-order polynomial fitting algorithm will be applied to the SPR signal processing to improve the sensitivity of the SPR chip, and it is also expected to apply the gain SPR chip to multi-component detection.

2. Materials and Methods

2.1. Materials

The experimental materials required for this experiment include: IR-140 dye was purchased from Xibao Biotechnology (Shanghai) Co., Ltd.; powdered PMMA with a relative molecular mass of 350,000 Mw was purchased from Xibao Biotechnology (Shanghai) Co., Ltd. Slides are purchased from Beijing Chemical Reagent Company, China.

2.2. Preparation

The instruments required for this experiment include: Electric heating constant temperature air drying oven, purchased from Beijing Ying'an Meicheng Scientific Instrument Co., Ltd.; multi-functional magnetic enhanced reactive ion etching machine, model ME-3A, purchased from the Microelectronics Center of the Chinese Academy of Sciences; cleaning station, spin coating instrument; The step meter for measuring the thickness of the gain medium layer was purchased from KLA-Tencor, USA, and its model was ASIQ; the sputtering system was purchased from Denton Vacuum Equipment (Beijing) Co., Ltd.

In this paper, the gain dielectric layer used in the gain SPR sensor chip is obtained by dissolving PMMA and IR-140 dye in methylene chloride. Then, a metal film is sputtered on the gain medium layer to complete the fabrication of the sensor chip. The specific process is as follows:

1) Cleaning the glass substrate: sequentially boiling the glass substrate with concentrated sulfuric acid and deionized water, and drying;

2) 200 mg of PMMA having a relative molecular mass of 35,0000 Mw was dissolved in 3 ml of dichloromethane; The IR-140 dye was added to the mixed solution at a specific gravity of 0.4%, 0.8%, 1.2%, and 1.6% to prepare different kinds of gain SPR chips. The amount of PMMA in this step can then be determined by the experimental conditions, but it is guaranteed to be completely soluble in the dichloromethane solution.

3) The gain dielectric layer is prepared by spin coating. The spin coating is carried out by first homogenizing at 500 rpm for 15 s and then spin coating at 5000 rpm for 1 min; then, the gain dielectric layer film is used. Bake at 100 degrees Celsius for five minutes;

4) Using a sputtering or deposition method, a gold film is prepared on the glass substrate obtained in the step (3), thereby obtaining a gain SPR chip;

2.3. Testing

The thickness of the gain dielectric layer of four SPR chips with different concentrations of IR-140 dye is measured by the step meter(KLA-Tencor, USA). The SPR chip was tested in a multi-channel image SPR analysis system developed by the laboratory itself [11].

3. Results and discussion

3.1. Design of the gain SPR sensor chip

Based on the literature and the existing research in this laboratory [12], this paper designs a three-layer SPR chip, as shown in Figure 1. The chip comprises three layers, namely a glass substrate, a gain layer and a metal layer. The chips in Figure 2 are the actual fabricated. Figure 2(a) is the chip prepared by the IR-140 dye having a specific gravity of 1.6%. Figure 2(b) is the chip prepared by the IR-140 dye having a specific gravity of 1.2%. Figure 2(c) is the chip prepared by the IR-140 dye having a specific gravity of 0.8%. Figure 2(d) is the chip prepared by the IR-140 dye having a specific gravity of 0.4%.

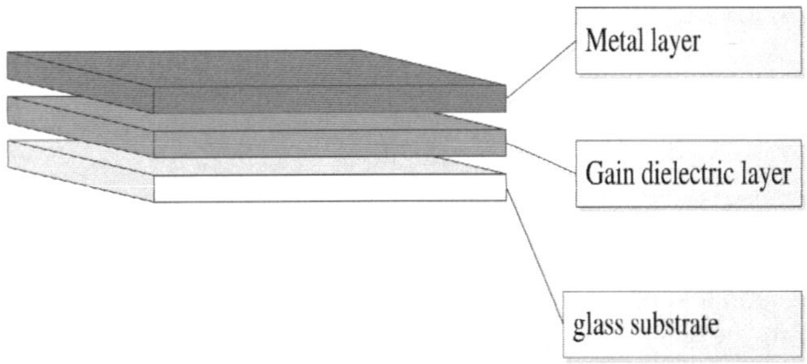

Figure 1. Gain SPR chip structure.

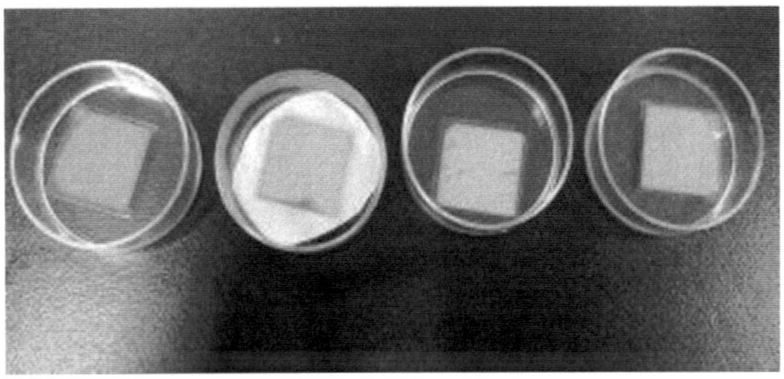

(a) (b) (c) (d)

Figure 2. Actually produced chip.

(a) The chip prepared by the IR-140 dye having a specific gravity of 1.6%.

(b) The chip prepared by the IR-140 dye having a specific gravity of 1.2%.

(c) The chip prepared by the IR-140 dye having a specific gravity of 0.8%.

(d) The chip prepared by the IR-140 dye having a specific gravity of 0.4%.

3.2. SPR sensor chip test

The thickness of the SPR chip in Figure 2 was tested and the SPR absorption peak was detected using the multi-channel image SPR biochemical analysis system of the laboratory. In this paper, the four different concentrations of IR-140 gain SPR chips were tested using our laboratory image SPR analysis system. The test is to install the gain SPR chip on the image SPR analyzer under normal temperature conditions and use deionized water as the analysis liquid layer to perform angle scanning and data acquisition.

The thickness of the gain dielectric layer of the SPR chip of Figure 2 was tested using a step meter (KLA-Tencor, USA), the thickness of which is shown in the following table:

Table 1. Gain level thickness.

IR-140 specific gravity (percentage)[a]	Gain dielectric layer thickness（um）
0.4%	3.5
0.8%	4.0
1.2%	5.0
1.6%	6.5-8

[a] percentage= IR-140 quality/ PMMA quality.

The SPR curve tested by the image SPR biochemical analysis system developed by our laboratory is shown in Fig. 3. In Figure 3, 0.4%, 0.8%, 1.2%, and 1.6% represent the specific gravity of IR-140 dye and PMMA, respectively. As can be seen from Figure 3, SPR characteristics have emerged.

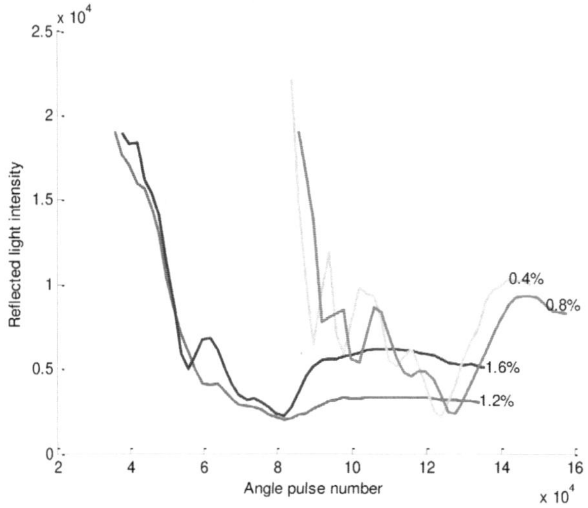

Figure 3. SPR raw test curve.

0.4%, 0.8%, 1.2%, and 1.6% represent the specific gravity of IR-140 dye and PMMA, respectively.

3.3. Optimize data processing parameters

Most SPR sensor data processing methods are based on the localization of the minimum SPR peak of the reflectance. These processing methods include nonlinear Lorentz curve method, centroid algorithm, optimized centroid algorithm and linear projection method, linear projection based on model parameters (MPLP) and 2-7 order polynomial fitting method. In this paper, we choose polynomial fitting for data processing. This is why the method is small in calculation and easy to implement, and it can also make the SPR curve smooth and reduce noise.

3.3.1. The mathematical model of the polynomial fitting algorithm. The mathematical model of the polynomial fitting algorithm is as follows:

$$f(x)=a_1x^m+\cdots+a_mx^x+a_{m+1} \qquad (1)$$

Here, m is the order of the polynomial, $a_1,\ldots,a_{m+1}$ is the fitting coefficient.

The most common method of solving the coefficients in equation (1) is the linear least squares method. The coefficient is determined as follows:

$$J(a_1,\ldots,a_{m+1})=\sum_{i=1}^{n}\delta_i^2=\sum_{i=1}^{n}\left[f(x_i)-y_i\right]^2 \qquad (2)$$

Where J is the fitting error sum, $f(x_i)$ represents the value of the fitted function at x_i , y_i representing the value corresponding to the original data.

In order to solve $a_1,\ldots,a_{m+1}$, Simply minimize the equation (2), which is $\frac{\partial J}{\partial a_k}=0\ (k=1,\ldots,m)$.

Get the equations for $a_1,\ldots,a_{m+1}$.

$$\sum_{k=1}^{m}a_k\left[\sum_{i=1}^{n}x_i^j\times x_i^k\right]=\sum_{i=1}^{n}x_i^j y_i,\ (j=1,\ldots,m) \qquad (3)$$

Assume

$$R=\begin{bmatrix} x_1 & \cdots & x_1^m \\ \vdots & \vdots & \vdots \\ x_n & \cdots & x_n^m \end{bmatrix}_{n\times m}$$

$$A=\left[a_1,\ldots,a_m\right]^T,\ Y=\left(y_1,\ldots,y_n\right)^T$$

Then equation (3) can be expressed as

$$R^T RA=R^T Y \qquad (4)$$

Therefore, the coefficient matrix A of the polynomial fitting can be obtained by the equation (4).

$$A=\left(R^T R\right)^{-1}R^T Y$$

In this paper, a prepared gain SPR chip with an IR-140 concentration of 0.4% was selected and tested using the image SPR biochemical analysis system of the laboratory to collect data. The fitting order of the selected polynomial is 2-11, and the fitting error is calculated by using equations (1) and (2), the fitting coefficient is calculated by equation (4). The fitting coefficient is shown in Table 2, where in Table 2, only the fitting coefficients of the 2nd to 5th order polynomials are given. And the fitting curve is drawn, as shown in Figure 4 and Figure 5. In Figure 4, Figure 4(a) is a fitting error curve, and Figure 4(b) is a fitting curve of 2-5 steps. In this paper, a 2-11 fitting error graph is made, but in this analysis, only the 2-5-order graph in Figure 4 is given, and the remaining fitting error graphs are in the appendix. 4 is an enlarged view of Figure 4(b). Referring to Figure 4(a) and Figure 5, the effect is ideal when the fitting order is 5 or more.

Table 2. Polynomial fitting coefficient table.

Fitting order	a_1	a_2	a_3	a_4	a_5	a_6
2	36914.632	-0.690	0			
3	78101.859	-2.362	0	0		
4	90026.686	-3.023	0	0	0	
5	-11033.421	4.070	-0.002	0	0	0

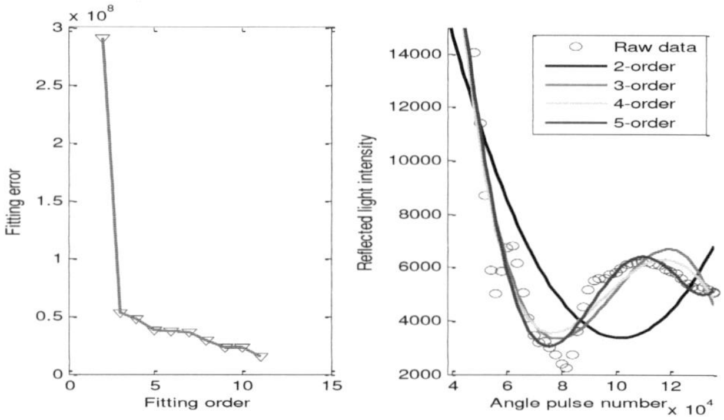

(a) (b)

Figure 4. Fitting error and fitting curve of prepared gain SPR chip with IR-140 concentration of 0.4%.

(a) fitting error curve (b) fitting curve of 2-5 steps.

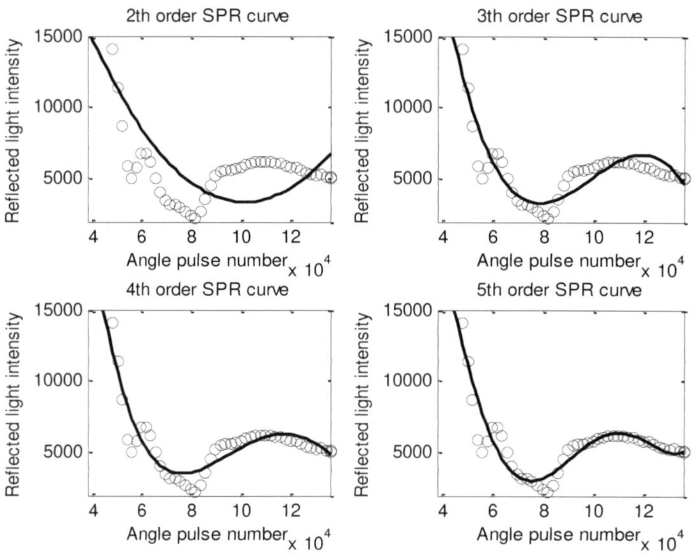

Figure 5. Magnified image of the prepared gain SPR chip with IR-140 concentration of 0.4%.

3.3.2. Determining the polynomial fitting order of multiple sets of SPR chips. From the above analysis, the fitting order of the polynomial is determined to be 5 or more. The fitting error calculations were performed on the four IR-140 concentration gain SPR chips and plotted. As shown in Figure 6. In Figure 6, it can be seen that for four different gain layer thickness SPR chips, when the fitting order is greater than 5, the fitting error does not change much, but it can be seen from the four curves. When the order number is greater than or equal to 9, the fitting error curve becomes more gradual. Considering the comprehensive consideration, this paper chooses the 9th-order polynomial fitting.

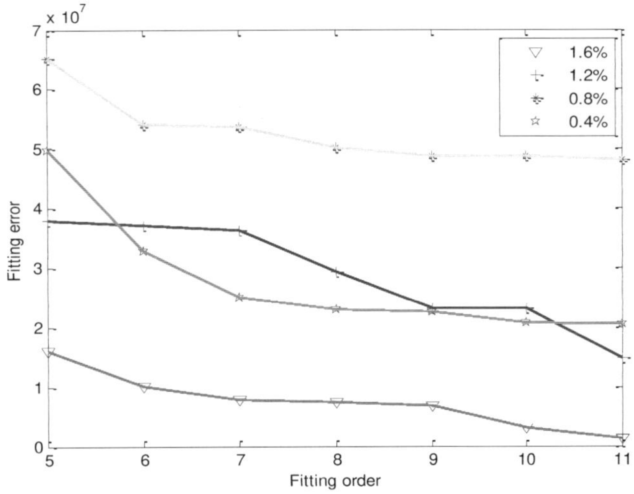

Figure 6. Four IR-140 concentration gain SPR chips for fitting error map.

0.4%, 0.8%, 1.2%, and 1.6% represent the specific gravity of IR-140 dye and PMMA, respectively.

3.4. SPR absorption peak curve under optimized conditions

In this paper, a layer of gain medium film was prepared by dissolving PMMA and IR-140 in 3 ml of dichloromethane. Based on the IR-140 concentration, four gain layers were fabricated and tested using our laboratory image SPR analysis system. Figure 7 is a graph of experimental results. Figure 7 shows the polynomial-fitted curve obtained by testing the gain SPR sensor chip prepared with four different concentrations of IR-140 dye; Figure 7 is normalized. The principle of normalization is: first select the lowest point of the SPR curve measured by the bare chip, that is, the SPR peak, and obtain its coordinates; then, based on this coordinate point, the four SPR curves in this experiment are also translated. That is normalized, so that its SPR peak is in the same position. It can be seen from Fig. 7 that the lower the IR-140 concentration, the smaller the half width of the prepared SPR sensor chip, and the higher the slope near the SPR peak. When the concentration of IR-140 is 1.6%, the half width is higher than the half width of 1.2%. It is speculated that the higher concentration of IR-140 may reduce the half width.

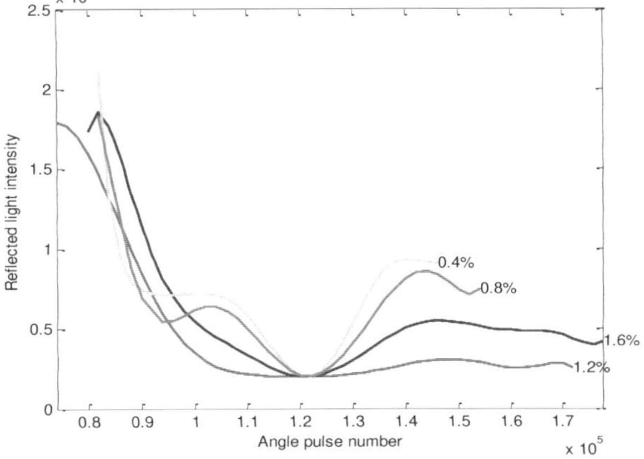

Figure 7. Nine-order polynomial fitting SPR absorption peak curve.

0.4%, 0.8%, 1.2%, and 1.6% represent the specific gravity of IR-140 dye and PMMA, respectively.

4. Summary

This paper describes in detail the working principle of the gain SPR sensor and the process flow of its fabrication. In order to compensate the loss caused by the SPPs excited by the SPR sensor, a layer of gain layer medium is added to the SPR chip by using the Kretschmann prism coupling structure. Experiments have shown that the SPR absorption peak of this gain-type SPR sensor will increase, that is, it will compensate for the SPPs loss. And this paper also gives an SPR signal processing algorithm based on 9th order polynomial fitting. During the fabrication of the gain SPR sensor, the spin-on speed and the ratio of PMMA to IR-140 dye have an important effect on the thickness of the gain layer. Next, we will delve into the characteristics of the gain dielectric layer. For example, the effect of the thickness of the gain dielectric layer on the SPR chip and the specific gravity of the IR-140 dye on the sensitivity of the chip.

5. Appendices

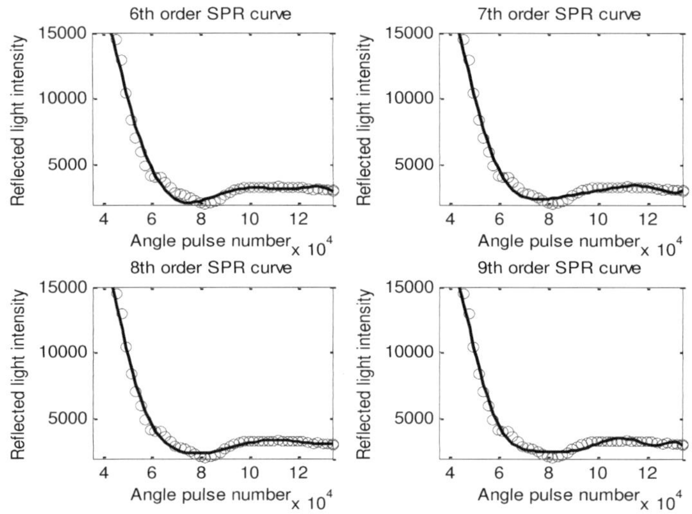

Figure 8. Magnified image of the prepared gain SPR chip with IR-140 concentration of 0.4%.

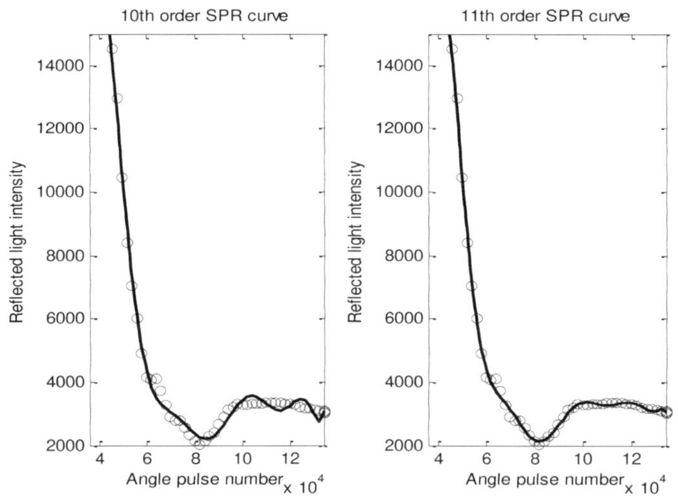

Figure 9. Magnified image of the prepared gain SPR chip with IR-140 concentration of 0.4%.

Acknowledgments
This work has been supported by the National High Technology Research, the National Basic Research Program of China (973 Program) (Grant No. 2014CB744600), and the National Natural Science Foundation of China (Grant Nos. 31571007, 61571420, 61372055).

References
[1] Siu P Ng, Yan Y Yip and Chi-Man L Wu 2015 *J. Sensors and Actuators B* **210** 36-45
[2] Ayda A, Hossein S, Hamidreza F and Mahmood S 2017 *J. IEEE SENSORS JOURNAL* **17** 4466-70
[3] Noginov M, Podolskiy V, Zhu G, Mayy M, Bahoura M, Adegoke J, Ritzo B and Reynolds K 2008 *J. OPTICS EXPRESS* **16** 1385-91
[4] Plotz G, Simon H and Tucciarone 1979 *J. Opt.Soc.Am* **69** 419-420
[5] Petr T, Jiri H 2005 *J. Sensors and Actuators B* **107** 162-69
[6] Garet G Nenninger, Marek P and Jiri H 2002 *J. Sci. Technol* **13** 2038-46
[7] Timothy M Chinowsky, Linda S Jung and Sinclair S Yee 1999 *J. Sensors and Actuators B* **54** 89-96
[8] Tiago A Sousa, Oliveira L, Loureiro F, Antonio M Lima and Neff H 2014 *J. IEEE* **1** 1-6
[9] Dhesingh R Shankaran, Gobi K and Norio M 2007 *J. Sensors and Actuators B* **121** 158-176
[10] Carlos A Filho, Antonio M and Nogueira L 2017 *J. IEEE SENSORS JOURNAL* **17** 6246-53
[11] Da F Cui, Hao Y Cai, Jun B Wang, Yang T Li, Jing Q Liang, Chang C Liu, Yu J Wang and Zi P Zheng 2007 *China. P* **200610066542.4** 2007-10-03
[12] Xing C, Lu L Zhang, Jian H Sun, Hui L and Da F Cui 2014 *J. Key Engineering Materials* **609** 430-434

IOP Conf. Series: Journal of Physics: Conf. Series **1209** (2019) 012007 doi:10.1088/1742-6596/1209/1/012007

Oven-controlled pressure sensor and its Mechanical aging

L D Du [1], Z Y Liu [1, 2], Z Fang [1, 2] and Z Zhao [1, 2]

[1] State Key Laboratory of Transducer Technology, Institute of Electronics, Chinese Academy of Sciences, Beijing, 100190, China
[2] University of Chinese Academy of Sciences, Beijing, 100090, China

The corresponding author's e-mail: lddu@mail.ie.ac.cn

Abstract. On the purpose of enhancing the performance of piezoresistive pressure sensor and reducing the difficulty of temperature compensation, the oven-controlled technology and mechanical aging method is studied in this paper. To be high efficiency and low consumption of oven-controlled method, the MEMS chip without package is used to reduce the thermal resistance. Furthermore, six-week mechanical aging was applied on the sensor. It is shown the mechanical aging can effectively improve the performance of pressure sensor. When the aginged sensor chip was tested at a range of -45℃ to 50℃, the designed sensor can measure the atmospheric pressure as an error less than ±0.2hPa with just consuming 0.8mW power.

1. Introduction

Usually, the high precision pressure sensor is developed with resonant principle, such as silicon resonant pressure sensor [1-5] and quartz resonant pressure sensor [6]. They cannot be widely used in handhold devices due to the high price and big volume. Piezoresistive pressure sensors have been massively used in electronic gadget due to the low price and small volume. However, to obtain high precision, piezoresistive pressure sensors will face the complicated compensation processes.

It is fortune that oven-controlled technology has been widely used to develop high performance oscillator due to the advantages of greatly reducing the difficulty of temperature compensation such as the oven-controlled FABR [7, 8], oven-controlled silicon crystal resonator [9], and oven-controlled quartz crystal oscillator [10]. All of them obtained good performance by full using of merit of oven-controlled technology. In order to take advantage of oven-controlled technology into pressure sensor, Reference [11] had developed an in-chip oven-controlled pressure sensor. However, it is easy affected by the leakage of the vacuum cavity. Reference [12] had also illustrated another oven-controlled pressure sensor. Although it has the characteristics of low error within ±0.5hPa, it is not good enough because the power consumption is relative high about 2.64W. Further work is needed to facilitate the application of the sensor.

For the development of oven-controlled sensor, the stability of sensor is demanded. Among the factors which affect the stability of sensors, mechanical aging will be the most common one. A lot of mechanical aging research focused on the material. In 2000, Michael D. Seale studied the effect of mechanical aging to composite stiffness [13]. And in 2015, Mohammad Hossein Sarfi used aging treatment to prepare experimental material [14]. In 2017, John A. Howarter analyzed the mechanical aging property of Ballistic Fibers [15]. Some mechanical aging research focused on sensor. In 1981, Yasuo Shin applied an USA patent about how to perform the aging treatment for semiconductor gas sensor. All of them illustrated the specifically processes to enhance the performance. When it comes to the mechanical aging of pressure sensor, there are few specific processes which were depicted. But it doesn't mean it isn't important vice versa it is very important so that every company keep it as the key commercial confidential information.

For the purpose of high efficiency and low consumption, the oven-controlled pressure sensor structure was designed and the mechanical aging was studied. The following content would be included in this paper. First of all, the whole sensor will be described. Secondly, the sensor will be simulated with the COMSOL software. Then the performance is tested. Subsequently, the mechanical aging of sensor was studied. The last not least, the sensor is verified by test results.

2. Structure and Simulation

The thermal structure of sensor was designed as that the MEMS chip was directly bonded on the PCB without package which is shown in Fig.1. The platinum resister (Pt 100) is just beneath and sticks closely with the MEMS ship. The heating resistors are located on the center of the print circuit board. Silicon grease filled between platinum resistors and heating resistors. The MEMS Chip was pasted on the PCB. The atmospheric pressure was exerted on the chip through the pipe of metal cap.

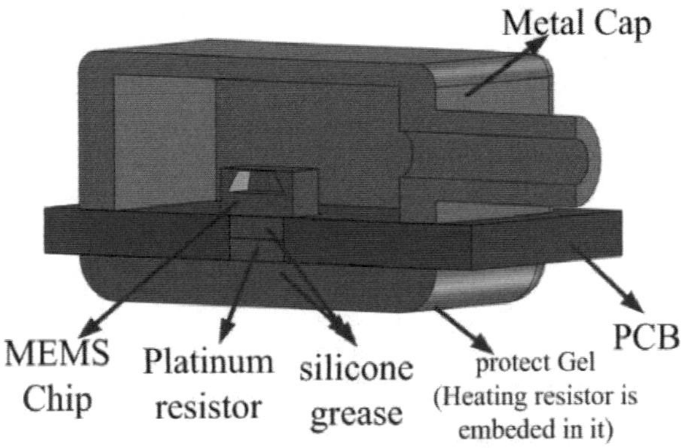

Figure 1. The thermal structure of sensor.

The principle of sensor is shown in Fig.2. The MEMS chip and Heater is electrically connected by microprocessor. By dynamically controlling the heating power the MEMS chip was kept running at the predefined value which was a little higher than the ambient temperature. The dynamically controlling was realized with the PWM module of microprocessor with a PID algorithm.

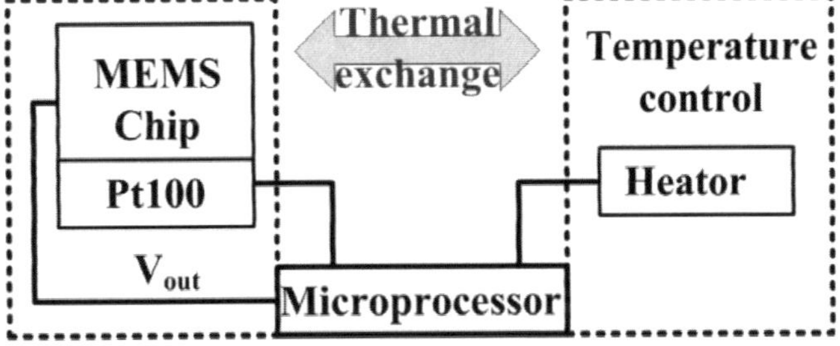

Figure 2. Principle of sensor.

The designed structure is simulated in COMSOL and shown in Fig. 3. In the paper, the ambient temperature was -50 ℃ and the pre-set temperature was 50 ℃. The metal cap, part of PCB and surface of silicon gel are set as the interface of heat convection. The parameters of steel, silicon, PCB and silicon gel come from the standard library of COMSOL software.

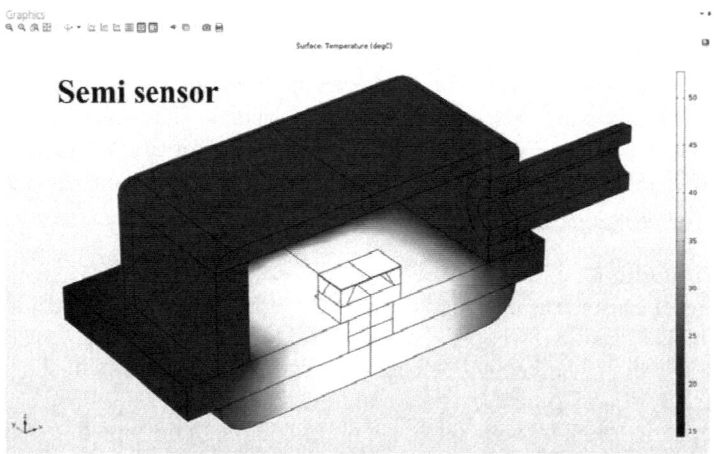

Figure 3. Simulation of sensor structure.

It can be obtained that the power consumption is just 0.1586 W to keep the sensor chip run at 50℃ when the ambient temperature is -50 ℃. It is shown as well that heating will not change the strain distribution of the silicon membrane.

3. Structure and Simulation

An evaluated sensor is developed as Fig.4.

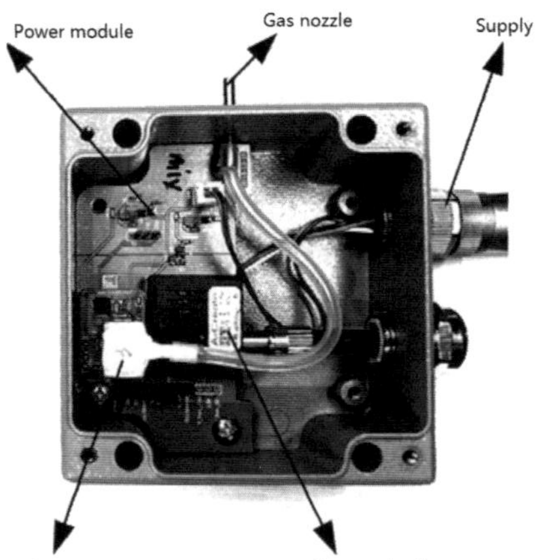

Figure 4. Developed sensor.

3.1. Power consumptions

The power consumptions are measured by oscilloscope shown in Fig. 5; and can be calculated by effective voltage and resistance of heater about 5Ω . It is shown in table 1 the maximum power consumption of whole system is about 0.8W. The duty cycle can be got from oscilloscope as well. It is suggested from the experiment data that the system can function well due to the maxim duty cycle just 27.1%; and there are enough driving capability for the circuit to maintain the sensor running in constant temperature 50℃.

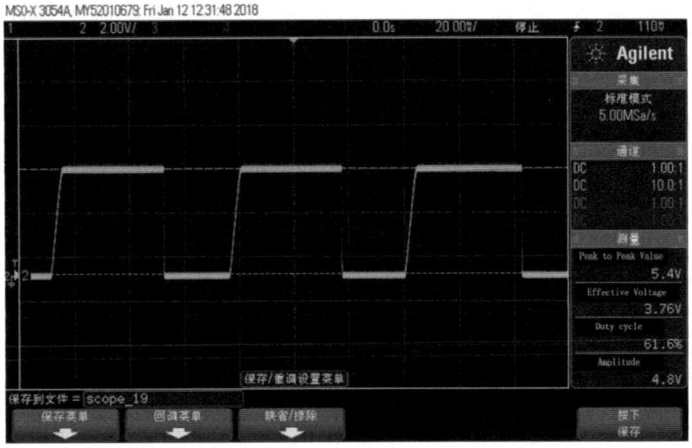

Figure 5. Results measured by oscilloscope.

Table 1. The power consumptions experiments data.

Ambient temperature (℃)	50	40	30	20	10	0	-10	-20	-30	-40	-50
Duty cycle (%)	0	5.3	8.3	10.6	13	15.8	17.9	20.3	23	25.5	27.1
Effective voltage (V)	0	0.78	1.01	1.18	1.32	1.47	1.59	1.71	1.82	1.93	2
Power consumption (W)	0	0.12	0.21	0.28	0.35	0.43	0.51	0.58	0.66	0.75	0.8

3.2. Aging

Six-week mechanical aging experiments are done for enhancing the performance of pressure sensor. During the aging, the pressurization-depressurization cycle was set about 10 minutes. And the sensors were measured once at a 6-day interval. The differences of the original output data between consecutive aging weeks are recorded and shown in Fig. 6. It is illustrated that the difference value of consecutive aging weeks is decreased when the aging time no more than five weeks while increased when the aging time longer than five weeks. It means that the sensor has an optimal aging time. To figure out this time is important for the application of sensor.

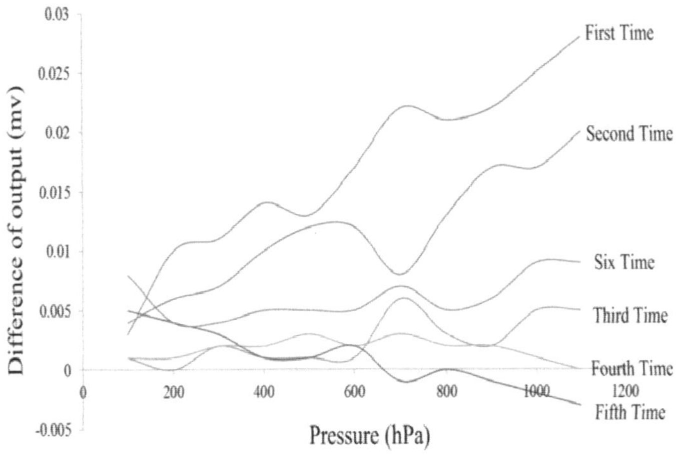

Figure 6. Results of Mechanical aging.

3.3. Performance of sensor

The aginged sensor was compensated at pre-set temperature and measured when the ambient temperature range from -45℃ to 45℃shown in Fig. 7(a) and (b). It is suggested in Fig. 7(b) that the developed sensor has a resolution within ±0.2hPa. Compared the designed sensor with reference [12] in table 2, the present sensor has better performance at resolution and power consumption.

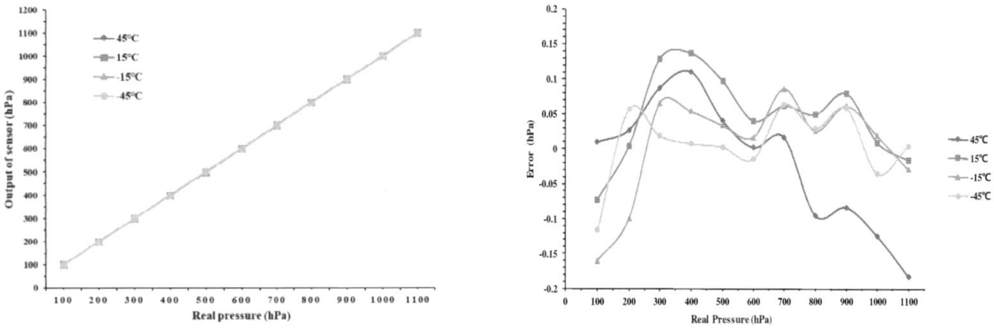

(a) (b)

Figure 7. Sensor's behaviour under different temperatures.

Table 2. Comparison the designed sensor with reference [12].

Reference	Full scale	Resolution	Power consumption
[14]	400~1100hPa	±0.5hPa	2.64W
This work	100~1100hPa	±0.2hPa	0.8W

4. Conclusions

In this paper, on the purpose of enhancing the performance of piezoresistive pressure sensor and reducing the difficulty of temperature compensation, the oven-controlled technology and mechanical aging method is studied. It is shown from the aging experiments that the sensor has an optimal aging time. To figure out this time is important for the application of sensor. When the aginged sensor chip was controlled at 50℃, it can measure the atmospheric pressure with an error less than ±0.2hPa by just calibrating it at ambient temperature 50℃. Such results manifested that the present sensor has better performance at resolution and power consumption.

5. Acknowledgments

This research was funded by National Natural Science Foundation of China (Grant 61871363) and by Key Project of Beijing Municipal Natural Science Foundation (Grant Z16002).

References

[1] Shi X Q, Lu Y L, Xie B, Li Y D, Wang J B, Chen D Y and Chen J, 2018 *Sensors* **18(8)** 2494
[2] Sadl M, Bradesko A, Belavic D, Bencan A, Malic B and Rojac T, 2018 *Sensors* **18(5)** 1423
[3] Wang L, Du X, Wang L, Xu Z, Zhang C and Gu D, 2017 *Sensors* **17(3)** 599
[4] Ren S, Yuan W, Qiao D, Deng J and Sun X, 2013 *Sensors* **13(12)** 17006-24
[5] Xie B, Xing Y, Wang Y, Chen J, Chen D Y and Wang J B, 2015 *Sensors* **15(9)** 24257-68
[6] Wang J B, Zhao C, Han D X, Jin X F, Zhang S M, Zou J B, Wang M M, Li W B. and Guo Y B, 2017 *Proceedings* **1(4)** 379

IOP Conf. Series: Journal of Physics: Conf. Series **1209** (2019) 012007 doi:10.1088/1742-6596/1209/1/012007

[7] Ruby R, Sankaragomathi K, Sridaran S and Parker R, 2015 *IEEE International Ultrasonics Symposium Proceedings* (Taipei, Taiwan) p 1-4
[8] Jabeom K, Kannan S, Ruby R and Otis B, 2017 *IEEE International Ultrasonics Symposium (IUS)*, (Washington D.C., USA)
[9] You W, Pei B, Sun K, Zhang L, Yang H and Li X X, 2017 *J. Micromech. Microeng* **(27)** 095002
[10] Huang Z G and Li J W, 2016 *Smart Science* **4(2)** 45–51
[11] Bruyker D D and Puers R, 2000 *Sens. Actuators, A* **82** 120–127
[12] Du L D, Zhao Z, Fang Z and Yan J Z, 2017 *IET Sci. Meas. Technol.* **11(7)** 907-913
[13] Michael D S and Eric I M, 2000 *J Compos Mater* **34(01)**
[14] Mohammad H S, Mohammad G and Arash B, 2015 *STEM Fellowship Journal* **1(2)** 21-25
[15] Howarter J A, Liu M, McDonough W G, Soles C and Holmes Ga A, 2017 *J. Polym. Sci. Pt. B-Polym. Phys.* **55** 1711-17

Electrospun nanofibrous membrane for electrodialysis

J X Jiang [1], G F Zheng [1], X Wang [2], G Fu [1], J Liu [1], W W Li [2], M F Zhuang [3] and SH M Guo [4]

[1] Department of Instrumental and Electrical Engineering, Xiamen University, Xiamen 361102, China
[2] School of Mechanical and Manufacturing Engineering, Xiamen Institute of Technology, Xiamen 361024, China
[3] School of Mathematical Sciences, Xiamen University, Xiamen 361021, China
[4] Fujian Provincial Key Laboratory of Mathematical Modeling and High Performance Scientific Computing, School of Mathematical Sciences, Xiamen University, Xiamen 361005, China

E-mail: zheng_gf@xmu.edu.cn

Abstract. The direct fabrication of homogenous membrane for electrodialysis has become the research hotspots in the field of water purification. A novel multi-jet electrospinning method with sheath gas was presented to study the fabrication of homogenous electrodialysis membrane. By soaking into sulfuric acid, the reactive exchange groups can be introduced into the electrospun nanofibrous membrane that has been hot pressed. Thanks to the pores among nanofibers, electrospun electrodialysis membrane had advantages of higher porosity, higher water uptake and lower membrane resistance, which promoted the ion transmission through the membrane. The porosity, water uptake and ion exchange capacity (IEC) reduced with the increase of membrane thickness. An electrodialysis separation system was set up to test its separation performance. The results showed that the desalinization ratio of NaCl solution could be reached to 58.15% within 30 minutes. The electrospun nanofibrous membrane for electrodialysis provides a simple and effective way to expand its industrial application.

1. Introduction

Electrodialysis [1-2] is a novel membrane separation method, which has shown great application potentials in the fields of water treatment, sea water desalination and medical industries [3-5]. However, the widely used preparation processes for separation membranes such as lamination are with complex procedures and high cost, which limits the further development of electrodialysis [6-7]. A simple process to prepare the homogenous membrane has been required urgently for the research of electrodialysis.

With the advantages of small diameter, high porosity and large specific surface area, electrospun nanofibers have been widely used in various fields, including air filtration [8], battery separator [9] and water treatment [10-11], which indicates a great application prospect for the rapid preparation of ion exchange membrane [12-13]. Pan [14] et al prepared cation exchange membranes by electrospinning and hot-pressing treatment, which showed approximately 4 times higher in the alkali dialysis coefficient compared with previous solution casting membrane. In addition, Pan's group [15] also fabricated anion exchange membrane by electrospinning for acid recovery, which exhibited both higher acid permeability and selectivity, underling the advantages of electrospun nanofibrous membrane in diffusion dialysis application. Electrodialysis is a technology based on the ion exchange membrane, then the fabrication of electrodialysis membrane by the method of electrospinning has been a potential way to improve its selective permeability further.

However, the low productivity of traditional electrospinning method with single-needle spinneret largely limits its industrial application. Multi-nozzle electrospinning is a straightforward method to improve the productivity of electrospun nanofibers, nevertheless there are mutual electrical interferences among the nozzles, which halts the increasing of charged jet density and the stable continuous ejection of charged jets [16-17]. Various auxiliary methods have been proposed to achieve multi-jet electrospinning with weakened interferences, such as the shield ring [18], the cylindrical electrode [19], the plate electrode [20], etc. How to fabricate uniform and compact nanofibrous membrane at a high efficiency has been the hotspot to expand its applications.

In this paper, a novel multi-nozzle spinneret with sheath gas was introduced into the fabrication of homogenous membrane for electrodialysis. The exchange performances of electrospun nanofibrous membrane were investigated and an electrodialysis separation system was set up to test its separation performance.

2. Materials and Methods

2.1. Preparation of electrospun membrane for electrodialysis

The schematic diagram of multi-jet electrospinning with the constraint of sheath gas to prepare electrospun nanofibrous membrane is shown in Figure 1, including a novel multi-jet spinneret, a precision syringe pump, a high voltage power source, an air compressor and a collector. The multiple nozzle tips of the spinneret were arranged in a triple layer equilateral hexagon configuration to achieve the high efficiency of electrospun nanofibers. Then, the assisted sheath gas was introduced into the spinneret as well. In our previous work, it has been reported that the sheath gas in laminar flow provided an additional constraining and stretching force to overcome the mutual interferences among the nozzles, decrease the critical voltage for multi-jet ejection, reduce the nanofiber diameter and promote the uniformity of electrospun nanofibrous membrane [21-23]. The design of the novel multi-jet electrospinning spinneret provides an available way to promote the simultaneous ejection of multiple jets and improve the nanofiber productivity.

To fabricate the electrospun nanofibrous membrane, the polyvinylidene fluoride (PVDF, Mw= 1,000,000 g/mol) powder was dissolved in the mixture solvent of N, N-Dimethylformamide and acetone with volume ratio of 5:5 (v:v). The mass concentration of the prepared solution was 10 wt%. A precision syringe pump (Pump 11 Pico Plus Elite, Harvard Apparatus America, USA) was used to supply polymer solution through the liquid inlet to the spinneret at the flow rate of 6,000 μl/h. Then, the solution could be supplied into each nozzle equally through the liquid cavity of the spinneret. A DC high voltage power source (DW-SA403-1ACE5, Dongwen high voltage power source Ltd. of Tianjin, China) was used to generate the high electrical field between the spinneret and the collector. The anode of the DC high voltage power source was connected to the nozzles of the spinneret, and the cathode was connected to the grounded collector of thin copper plate. The applied voltage was set to be 16 kV. An air compressor was used to pump the assisted gas into the spinneret through the gas inlet with the gas pressure of 25 kPa. The distance between the lowest nozzle tip and the collector was 8 cm. The collector was fixed on the XY motion platform (GXY1515GT4, Googoltech, China) to guide the electrospun nanofibers deposited into a uniform membrane.

After the electrospinning process, the membrane was hot-pressed at the temperature of 80℃ to promote its mechanical strength. Then, to prepare the electrodialysis membrane with selective permeability, the electrospun membrane was soaked into 98% concentrated sulfuric acid for 12 hours at the temperature of 60℃ to introduce the reactive exchange groups into membrane. Next, the membrane was cleaned with 60% sulfuric acid, 30% sulfuric acid and deionized water, successively to remove the sulfuric acid at the membrane surface. Finally, the prepared electrodialysis membrane was kept at the deionized water for further testing. In the experiments, the micro/nano morphologies of the membrane were observed and recorded by a field emission scanning electron microscope (SEM, Hitachi SU-70, Japan).

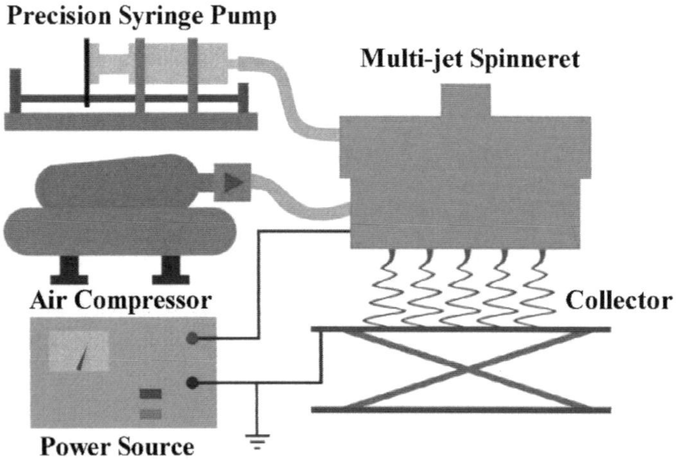

Figure 1. Schematic diagram of multi-jet electrospinning with the constraint of sheath gas.

2.2. Characterization of electrospun membrane for electrodialysis

2.2.1. Porosity. The porosity of membrane was measured according to the n-butanol uptake method. The electrodialysis membrane was cut into the square shape with each side length of 2 cm. Then, it was thrown into n-butanol to fill with the holes inside the membrane. 2 hours later, it was taken out from the n-butanol and the residual solution was removed slightly. The porosity of the membrane could be calculated as [24]:

$$\varepsilon = \frac{m_2 - m_1}{\rho A h} \tag{1}$$

Where, m_1 and m_2 represented the mass of membrane before and after soaked into n-butanol, ρ was the density of n-butanol, A was the area of membrane, h was the thickness of membrane.

2.2.2. Water uptake. The water uptake of membrane was an index that performed the water content inside the membrane. The membrane of square shape with each side length of 2 cm was thrown into the deionized water for 24 hours. Then, it was dried in a vacuum oven at the temperature of 80 ℃ to remove the water completely. Then, the water uptake of membrane could be calculated as::

$$C_W = \frac{w_1 - w_2}{w_1} \times 100\% \tag{2}$$

Where, w_1 and w_2 were the mass of wet membrane and dry membrane, respectively.

2.2.3. Ion exchange capacity (IEC). The ion exchange capacity (IEC) was measured according to the conventional Mohr method. The membrane was dried in a vacuum oven for 2 hours. Then, it was thrown into the NaCl solution was concentration of 0.5 mol/L for 24 hours, so that the H^+ inside the membrane and the Na^+ in the solution could be transferred fully. Finally, the released amount of H^+ was determined through titration with 0.01 mol/L NaOH solution. The value of IEC could be calculated as [25]:

$$IEC = \frac{V \times c}{m} \tag{3}$$

Where, V was the volume of NaOH solution used for titration, c was the concentration of NaOH solution, m was the mass of dry membrane.

3. Results and Discussion

The morphology of electrospun nanofibrous membrane for electrodialysis was investigated first. The SEM images of electrodialysis membrane fabricated by lamination and electrospinning are shown in Figure 2a and Figure 2b, respectively. Compared with most commercial membrane fabricated by complex processes such as lamination, the electrospun membrane was composed of a large number of nanofibers and there existed several pores among nanofibers, which increased the contact area between membrane and solution, contributing to the promotion of ion transmission. After soaked into 98% concentrated sulfuric acid, the surface morphology of nanofibers became rougher and waver with the average diameter increasing from 0.68 μm to 0.83 μm. Due to the water swelling of active groups, there would be some small granular structures on the surface of nanofibers and some nanofibers would gather together, as shown in Figure 2c.

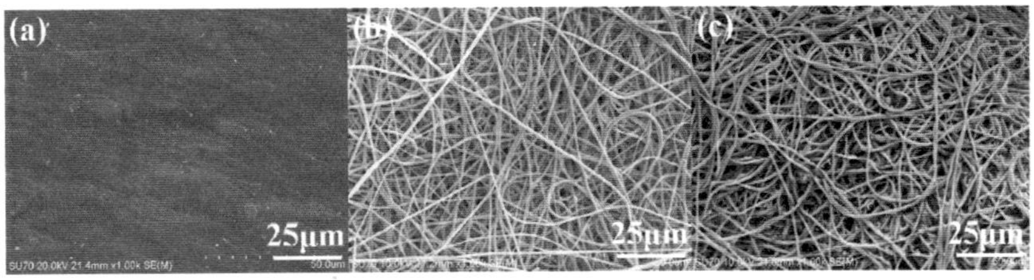

Figure 2. SEM images of electrodialysis membrane fabricated by lamination and electrospinning.

The performances of electrodialysis membrane fabricated by lamination and electrospinning were tested, as given in Table 1. It could be concluded that the electrospun membrane has a larger porosity and water uptake due to the large specific area inside the membrane. However, limited by the present reaction condition, ionic exchange capacity (IEC) for the electrospun membrane was a bit smaller, which could be enhanced by optimizing the post-processing of membrane and introducing more reactive exchange groups. As the nanofiber deposited layer by layer and the thickness of nanofibrous membrane increased, the porosity, water uptake and IEC of membrane decreased. When the average thickness of electrospun membrane increased from 43 μm to 164 μm, the porosity, water uptake and IEC reduced from 79.8% to 64.6%, 65.3% to 62.2% and 0.507 mmol/g to 0.370 mmol/g, respectively. Attributing to the larger porosity and water uptake, there were more ion transmission passageways inside the electrospun membrane, resulting in a much smaller membrane resistance than that of the membrane fabricated by lamination, which was beneficial for the ion exchanging and the energy consumption reducing. In Table 1, sample 1 is an electrodialysis membrane fabricated by lamination, while other samples are all electrodialysis membranes fabricated by electrospinning.

Table 1. Exchange performances of electrodialysis membrane.

	Thickness (μm)	Porosity (%)	Water uptake (%)	IEC (mmol/g)
1	44	39.1	40.1	0.707
2	43	79.8	65.3	0.507
3	49	79.5	64.8	0.504
4	94	71.2	64.9	0.431
5	103	72.2	64.2	0.430
6	152	68.1	63.4	0.385
7	164	64.6	62.2	0.370

Then, a simple experimental setup was established to test the electrodialysis performance of electrospun nanofibrous membrane, as illustrated in Figure 3, including an anode chamber, a dilute chamber and a cathode chamber. An electrospun nanofibrous membrane of which the cation exchange groups had already been introduced was fixed between the cathode chamber and the dilute chamber, while an anion-exchange membrane was fixed at another side to transmit the ions in the solution selectively. A 30 V direct-current voltage was supplied between the anode and the cathode to provide a driving force for ion transmission. Here, 0.025 mol/L NaCl solution was utilized to investigate the ion separation behavior for electrospun nanofibrous membrane. The varieties of solution conductivity in each chamber were recorded as time going by as depicted in Figure 4. The solution conductivity in the dilute chamber decreased from 3.8 mS/cm to 1.59 mS/cm in 30 minutes while those in the anode and cathode chambers increased, which indicated that the ions had been separated selectively, verifying the feasibility of the electrospun nanofibrous membrane in the application of electrodialysis.

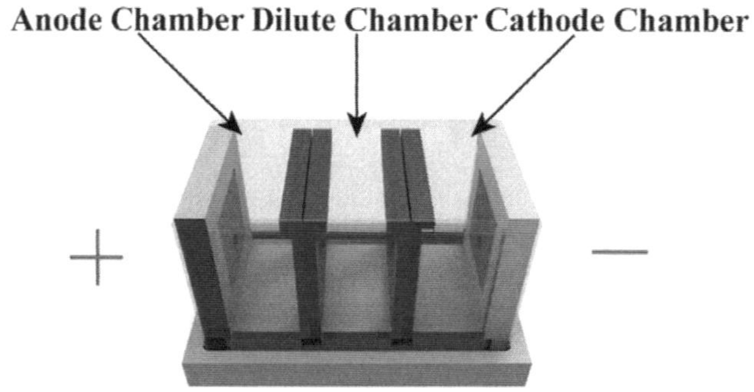

Figure 3. Experimental setup for electrodialysis performance.

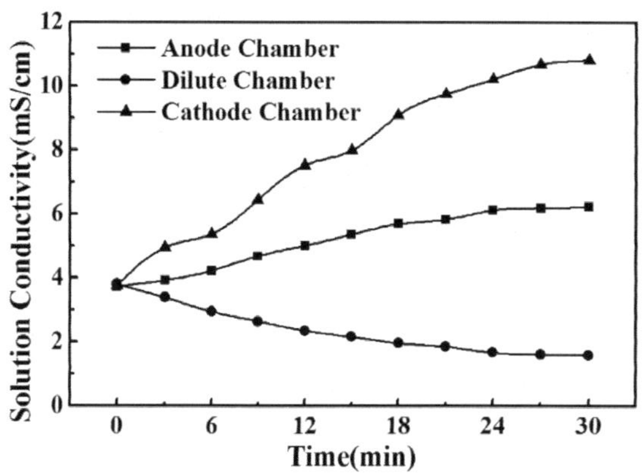

Figure 4. Relationship between NaCl solution conductivity and testing time with different chambers.

4. Conclusions

The technology of multi-jet electrospinning is introduced into the fabrication of electrodialysis membrane. To promote the simultaneous ejection of multiple jets, a novel multi-jet spinneret with the constraint of sheath gas was proposed, thus uniform nanofibrous membrane could be obtained at a high productivity. Then, through the processes of hot-pressing, introducing the reactive exchange groups and removing the surface sulfuric acid, electrospun membrane for electrodialysis was achieved successfully. Compared with electrodialysis membrane fabricated by lamination, the electrospun membrane for electrodialysis presented a higher porosity, a higher water uptake and a lower membrane resistance. From the membrane electrodialysis experiments, it could be seen that the desalinization ratio of NaCl solution could be reached to 58.15% within 30 minutes, exhibiting an excellent performance in electrodialysis for electrospun nanofibrous membrane.

Acknowledgments
This research was financially supported by the National Natural Science Foundation of China (No. 51575464, 51805460), Science and Technology Planning Project of Fujian Province (No. 2017H0037, 2017J01099), Education and Scientific Research Projects for Middle-aged and Young Teachers of Fujian Province of China (No. JAT170816).

References

[1] Xu T and Huang C 2008 Electrodialysis-based separation technologies: A critical review *Aiche J.* **54** 3147–59
[2] Campione A, Gurreri L, Ciofalo M, Micale G, Tamburini A and Cipollina A 2018 Electrodialysis for water desalination: A critical assessment of recent developments on process fundamentals, models and applications *Desalination* **434** 121-60
[3] Galama A H, Daubaras G, Burheim O S, Rijnaarts H H M and Post J W 2014 Seawater electrodialysis with preferential removal of divalent ions *J. Membrane Sci,* **452** 219-28
[4] Xu X, Lin L, Ma G, Wang H, Jiang W, He Q, Nirmalakhandan N and Xu P 2018 Study of Polyethyleneimine Coating on Membrane Permselectivity and Desalination Performance during Pilot-scale Electrodialysis of Reverse Osmosis Concentrate *Sep. Purif. Technol.* **207** 396-405
[5] Jones R J, Massanetnicolau J, Guwy A, Premier G C, Dinsdale R M and Reilly M 2015 Removal and recovery of inhibitory volatile fatty acids from mixed acid fermentations by conventional electrodialysis *Bioresource Technol.* **189** 279-84
[6] Venugopal K and Dharmalingam S 2013 Fundamental studies on a new series of SPSEBS - PVA - QPSEBS bipolar membrane: Membrane preparation and characterization *J. Appl. Polym. Sci.* **127** 4983-90
[7] Xie M, Shon H K, Gray S R and Elimelech M 2016 Membrane-based processes for wastewater nutrient recovery: Technology, challenges, and future direction *Water Res,* **89** 210-21
[8] Li J, Zhang D, Yang T, Yang S, Yang X and Zhu H 2018 Nanofibrous Membrane of Graphene Oxide-in-Polyacrylonitrile Composite with Low Filtration Resistance for the Effective Capture of PM 2.5 *J. Membrane Sci,* **551** 85-92
[9] Zhu P, et al. 2017 A novel bi-functional double-layer rGO–PVDF/PVDF composite nanofiber membrane separator with enhanced thermal stability and effective polysulfide inhibition for high-performance lithium–sulfur batteries *J. Mater, Chem. A* **5** 15096-104
[10] Ma W, et al. 2018 Durable superhydrophobic and superoleophilic electrospun nanofibrous membrane for oil-water emulsion separation *J. Colloid Interf. Sci.* **532** 12-23
[11] Wu R X, Zheng G F, Li W W, Zhong L B and Zheng Y M 2018 Electrospun Chitosan Nanofiber Membrane for Adsorption of Cu(II) from Aqueous Solution: Fabrication, Characterization and Performance *J. Nanosci. Nanotechno.* **18** 5624-35
[12] Chitpong N and Husson S M 2017 High-capacity, nanofiber-based ion-exchange membranes for the selective recovery of heavy metals from impaired waters *Sep. Purif. Technol.* **179** 94-103
[13] DeGostin M B, Peracchio A A, Myles T D, Cassenti B N and Chiu W K S 2016 Charge transport in the electrospun nanofiber composite membrane's three-dimensional fibrous structure *J. Power Sources* **307** 538-51
[14] Pan J, Ge L, Lin X, Wu L, Wu B and Xu T 2014 Cation exchange membranes from hot-pressed electrospun sulfonated poly(phenylene oxide) nanofibers for alkali recovery *J. Membrane Sci,* **470** 479-85

[15] Pan J, He Y, Wu L, Jiang C, Wu B, Mondal A N, Cheng C and Xu T 2015 Anion exchange membranes from hot-pressed electrospun QPPO–SiO 2 hybrid nanofibers for acid recovery *J. Membrane Sci,* **480** 115-21

[16] Li W, Zheng G, Wang X, Zhang Y, Li L, Wang L, Wang H and Sun D 2011 Directly electrospun ultrafine nanofibres with Cu grid spinneret *J. Phys. D. Appl. Phys.* **44** 135502

[17] Theron S A, Yarin A L, Zussman E and Kroll E 2005 Multiple jets in electrospinning: experiment and modeling *Polymer* **46** 2889-99

[18] Yang Y, Jia Z, Li Q, Hou L, Liu J, Wang L, Guan Z and Zahn M 2010 A shield ring enhanced equilateral hexagon distributed multi-needle electrospinning spinneret *IEEE T. Dielect. El. In.* **17** 1592-601

[19] Kim G H, Cho Y S and Wan D K 2006 Stability analysis for multi-jets electrospinning process modified with a cylindrical electrode *Eup. Polym. J.* **42** 2031-8

[20] Zheng Y, Zhuang C, Gong R H and Zeng Y 2014 Electric Field Design for Multijet Electropsinning with Uniform Electric Field *Ind. Eng. Chem. Res.* **53** 14876-84

[21] Zhao Y, Jiang J, Li W, Wang X, Zhang K, Zhu P and Zheng G 2016 Electrospinning jet behaviors under the constraints of a sheath gas *Aip Adv.* **6** 167-628

[22] Zhao Y, Jiang J-X, Chen D-Y, Zheng G-F and Sun D-H 2015 High efficiency multi-jet electrospraying of nano particles with assisted gas *Optics and Precision Engineering* **23** 1062-9

[23] Yu Z J, Lin Y H, Huang W W, Zhuang M F, Hong Y Q, Zheng G F, Sun D H and Ieee, Multi Spinnerets Electrospinning with Assistant Sheath Gas. In *2014 9th Ieee International Conference on Nano/Micro Engineered and Molecular Systems*, Ieee: New York, 2014; pp 64-7.

[24] Chen W, Liu Y, Ma Y and Yang W 2015 Improved performance of lithium ion battery separator enabled by co-electrospinnig polyimide/poly(vinylidene fluoride-co-hexafluoropropylene) and the incorporation of TiO 2 -(2-hydroxyethyl methacrylate) *J. Power Sources* **273** 1127-35

[25] Jang W G, Hou J and Byun H s 2011 Preparation and characterization of PVdF nanofiber ion exchange membrane for the PEMFC application *Desalin. Water Treat.* **34** 315-20

A carpet cloak based on gradient metasurfaces

R Wang[1], S Yue[1], Y Hou[1], S Liu[1], M Li[1] and Z Zhang[1,2]

[1]Microelectronics Instruments and Equipments R&D Center, Institute of Microelectronics of Chinese Academy of Sciences, Beijing, 100029, China

[2]Author to whom any correspondence should be addressed: zz241@ime.ac.cn

Abstract. In this paper, we numerically demonstrate a carpet cloak operating in reflection with the incident wavelength of 800 nm. The carpet cloak comprises gradient metasurfaces with a metal-insulator-metal configuration. The gradient metasurfaces reroute the indicent light and restore the reflectted wavefront by compensating the phase differences between each unit cell of the metasurface. The proposed cloak is ultrathin, simple and easy for practical implementation.

1. Introduction

Metamaterials are artificial composites made up of micro-structures at the subwavelength scale [1]. Due to their amazing ability to manipulate electromagnetic waves, metamaterials exhibit novel promising applications [1], such as invisibility cloaks, negative index of refraction, superlenses, superabsorber, and so on. Among all the applications above, the invisibility cloaks, which can hide objects from the detection of incident electromagnetic waves, have received considerable attention in recent years [2]. The most popular approaches for invisibility cloaks based on metamaterials are transformation optics [3] and scattering cancellation [4]. However, inherent issues, such as anisotropy, inhomogeneity, and complex electromagnetic parameters of metamaterials, make the fabrication of the device rather complicated. On the other hand, based on the quasi-conformal mapping technique [5], the carpet cloak can relax the requirements of anisotropy and material properties. However, they are still bulky in size, and it is a great challenge to realize an ultrathin carpet cloak.

The development of metasurfaces [6] provides a novel method to manipulate the optical waves. They consist of subwavelength-sized elements in an optically thin layer, and can locally tailor the response of light at the nanoscale. Metasurfaces enable a series of unique phenomena and applications, such as the anomalous reflections/refractions [7], planar optical lenses [8], and high-resolution hograms [9]. An ultrathin directional carpet cloak was demonstrated in the microwave region based on metasurfaces [10]. As for the cloaking in the regions of the near-infrared and visible light, an ultrathin invisibility skin cloak was proposed to conceal a 3D arbitrarily shaped object by completely restroating the relfected wavefront at the incident wavelength of 730 nm [11]. However, since the metasurfaces for the skin cloak tightly wrapped over the object, point-to-point design and fabrication were necessary for this nanostructue, rendering the cloaking scheme unrealistic.

In this letter, we numerically demonstrate a reflective carpet cloak under the incident wavelength of 800 nm. The carpet cloak consists of gradient metasurfaces with a metal-insulator-metal (MIM) configuration. A super cell of the metasurfaces comprises 5 units with different sizes of the top gold rods. The gradient metasurfaces will compensate the phase differences between the adjacent unit cells, and the reflected wavefront can be effectively restructured for a specific polarization of the incident beam. Two mirror sysmetrical gradient metasurfaces provide a triangular space, which is suitable to conceal arbitrarily shaped objects underneath. Point-to-point design and fabrication are not needed according to our configuration. The proposed carpet cloak is ultrathin, simple, and more straightforward for practical implementation.

2. Simulation Results and Discussions

The fundamental unit cell of the MIM metasurface is plotted in Figure 1(a). It comprises a gold rod (the thickness is t_1 = 30 nm) and an optically thick gold substrate (the thickness is t_3 = 150 nm) separated by a SiO_2 layer with the thickness of t_2 = 50 nm. The length and the width of the gold rod are L and w, and the subwavelength lateral dimentions of the unit cell are L_x = 120 nm and L_y = 300 nm, respectively. When a y-polarized incident beam (the vacuum wavelength is λ = 800 nm) normally illuminates the MIM metasurface, electric currents will be induced both in the gold rod and substrate. Since the gold rod and substrate are separated by a thin dielectric layer (the thickness of SiO_2 is only 50 nm), strong near-field coupling can be effectively excited. The induced antiparallel electric currents on the upper and lower gold layers lead to a magnetic resonance inside the intermediate dielectric layer at a particular frequency. Obviously, the magnetic resonance properties of the unit cell, such as the reflection amplitude and phase delay, can be efficiently tuned by the geometrical parameters of the nanostructures.

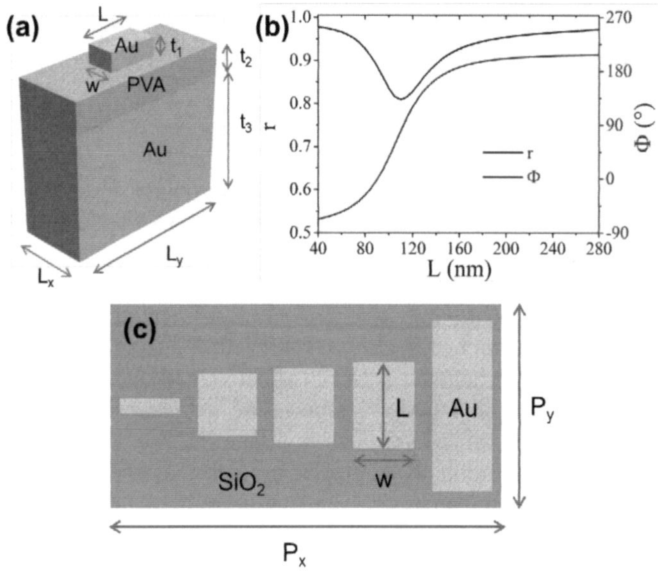

Figure 1. (a) The unit cell of MIM meatsurface with t_1 = 30 nm, t_2 = 50 nm, t_3 = 150 nm, L_x = 120 nm, L_y = 300 nm, and w = 90 nm. The length L of the upper gold rod varies. (b) The reflection amplitude (r) and phase (Φ) versus the length L of the upper gold rod. (c) Top view of the super cell of the gradient metasurface with $P_x = 5L_x$ = 600 nm and $P_y = L_y$ = 300 nm.

To verify the analysis above, the reflection characteristics of the unit cell are simulated by COMSOL Multiphysics. The width w of the upper gold rods is fixed as 90 nm. The refractive index of the SiO_2 layer is set to be n = 1.45, and the permittivity of gold is taken from the experiment results [12]. The reflection amplitude (r) and phase (Φ) as functions of the length L of the upper gold rod are illustrated by the black and blue curves in Figure 1(b). The reflection amplitude maintains a relatively high stable value (> 0.8), and does not vary too much according to the length L. Due to the Ohmic loss and the resonant absorption of the gap surface plasmons (GSPs), the reflection amplitude cannot reach 1. The reflection phase can be adjusted to a certain value according to the length L. Based on the simulation above, five different lengths (L = 20, 90, 110, 125, and 250 nm, respectively) of the upper Au rods are carefully selected to form a super cell of the gradient metasurfaces, the top view of which is plotted in Figure 1(c). The corresponding reflection phases provide a linear increase, as listed in Table 1. The gradient metasurfaces essentially support the anomalous reflection with high efficiency. The linear phase gradient can be written as

$$\Delta\varphi = \frac{2\pi}{P_x} = \frac{2\pi}{m \cdot L_x} \tag{1}$$

where $m = 5$ is the number of the unit cells in a super cell of gradient metasurfaces.

Table 1. The selected lengths and the corresponding reflection phases of the gradient metasurfaces.

The number of unit cells	L (nm)	Φ (°)
1	20	-72.5
2	90	-2.1
3	110	70.4
4	125	136.1
6	250	206.2

The operation of the carpet cloak is illustrated in Figure 2. The gradient metasurface is laid on the ground plane with a tilt angle of β. When the incident beam (the black solid arrows) normally illuminates the dielectric (SiO_2) and metallic (gold) layers without any metasurfaces (the upper gold rods), the reflected beam (the black dashed arrow) will be deflected towards an angle of 2β away from the incident direction. By introducing the gradient metasurfaces above the SiO_2 layer, the reflected beam can be steered back to the direction of the incident beam due to the anomalous reflection discussed above, as plotted by the red solid arrows.

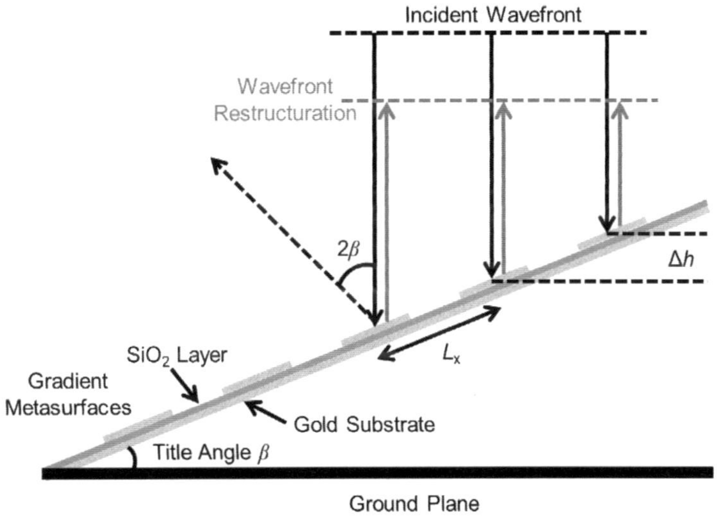

Figure 2. Schematic view of the carpet cloak based on gradient metasurfaces.

The optical path difference Δh between the adjacent unit cells of metasurfaces is

$$\Delta h = L_x \cdot \sin\beta \tag{2}$$

And the phase difference $\Delta\varphi$ between the adjacent unit cells is

$$\Delta\varphi' = 2k_0 \cdot \Delta h = 2 \cdot (\frac{2\pi}{\lambda_0}) \cdot L_x \cdot \sin\beta \tag{3}$$

where k_0 is the wave number in the free space, and L_x is the lateral distance between the adjacent unit cells along x axis. If the phase gradient provided by the metasurfaces can compensate the phase differences between the adjacent unit cells ($\Delta\varphi=\Delta\varphi'$), the reflected wavefront can be effectively resctructured as if it were reflected from the ground plane. The objects under the metasurfaces are concelled from the outside detection. Combining the equation (1) and equation (3), the tilt angle β of the carpet cloak is

$$\sin\beta=\frac{\lambda_0}{2\cdot P_x} \tag{4}$$

Accroding to the geometrical and optical parameters of the MIM gradient metasurfaces above, the tilt angle β of the carpet cloak can be calculated as 46.5°.

Based on the analysis and the design above, the cloaking characteristics of the gradient metasurfaces are simulated as well. The metasurface is horizontally laid, and the y-polarized incident beam (λ = 800 nm) illuminates the metasurfaces with an incident angle of 46.5°, corresponding to the tilt angle β = 46.5° of the carpet cloak, as plotted in Figure 3(a). Periodic boudary conditions are set to simulate the infinite carpet cloak and the plane wave excitation. As we can see from Figure 3(b), the wavefronts of the reflected E_y far field from the gradient metasurface are nearly parallel to the incident wavefronts in Figure 3(a), which means the reflected wavefronts are effectively restructured. The white solid arrows in Figure 3(a) and (b) illustrate the directions of the wave vectors of the incident and reflected light. Arbitrarily shaped objects can be hidden under the triangular space formed by two mirror sysmetrical gradient metasurfaces. The overall thickness of the metasurface (t_{total} = 230 nm) is only a quarter of the operating wavelength (λ = 800 nm), leading to an ultrathin cloaking device. The quality of the reflected plane wave is deteriorated by the fluctuation of the refelction amplitude r within a super cell due to the Ohmic loss and the resonant absorption of the gap surface plasmons. To improve the efficiency of the wavefront restructuration, dielectric metasurfaces [13] will be good candidates. Additionally, the polarization dependence of the device can be overcome by implementing a circularly symmetric cross section of unit cells [8].

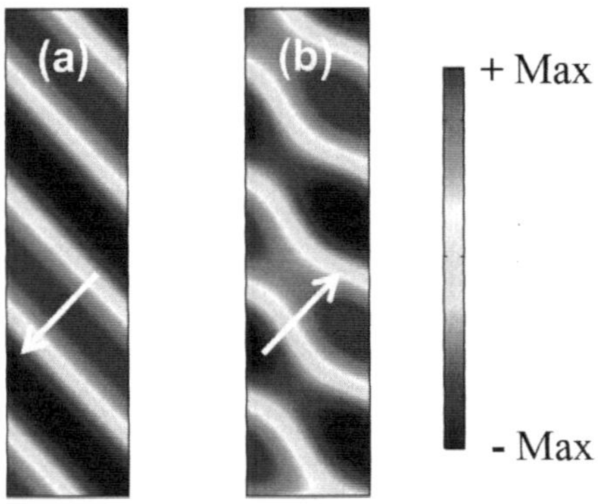

Figure 3. The simulated cloaking characteristics of the gradient metasurfaces. (a) The incident and (b) the reflected E_y far field patterns of the gradient metasurfaces.

3. Conclusion

In this paper, we have reported an ultrathin carpet cloak based on the gradient MIM metasurfaces. The gradient metasurfaces compensated the phase differences between the adjacent unit cells, and the reflected wavefronts were nearly reconstructed to be plane waves. Arbitrarily shaped objects could be hidden under the carpet cloak. The proposed carpet cloak is ultrathin, simple and easy for practical implementation.

4. References

[1] Liu Y and Zhang X 2011 *Chem. Soc. Rev.* **40**, 2494
[2] Schurig D, Mock J J, Justice B J, Cummer S A, Pendry J B, Starr A F and Smith D R 2006 *Science* **314**, 977
[3] Cai W, Chettiar U K, Kildishev A V and Shalaev V M 2007 *Nat. Photonics* **1**, 224
[4] Fan P, Chettiar U K, Cao L, Afshinmanesh F, Engheta N and Brongersma M L 2012 *Nat. Photonics* **6**, 380
[5] Li J and Pendry J B 2008 *Phys. Rev. Lett.* **101**, 203901
[6] Ni X, Emani N K, Kildishev A V, Boltasseva A and Shalaev V M 2012 *Science* **335**, 427
[7] Yu N, Genevet P, Kats M A, Aieta F, Tetienne J P, Capasso F and Gaburro Z 2011 *Science* **334**, 333
[8] Khorasaninejad M, Zhu A Y, Roques-Carmes C, Chen W T, Oh J, Mishra I, Devlin R C and Capasso F 2016 *Nano Lett.* **16**, 7229
[9] Huang L, Chen X, Bai B, Tan Q, Jin G, Zentgraf T and Zhang S 2013 *Light Sci. Appl.* **2**, e70
[10] Zhang J, Mei Z L, Zhang W R, Yang F and Cui T J 2013 *Appl. Phys. Lett.* **103**, 151115
[11] Ni X, Zi J W, Mrejen M, Wang Y and Zhang X 2015 *Science* **349**, 1310
[12] Johnson P B 1972 *Phys. Rev. B* **6**, 4370
[13] Hsu L Y, Lepetit T and Kanté B 2015 *Prog. Electromagn. Res.* **152**, 33

Acknowledgments

Project supported by State Key Laboratory of Applied Optics, Changchun Institute of Optics, Fine Mechanics and Physics (CIOMP), Chinese Academy of Sciences (CAS); CAS Pioneer Hundred Talents Program.

Research on thick copper CMP based on multilayer wiring of radio frequency microsystem

K Q Qian[1], H Z Zhang[1], L Wang[1], D R Wang[1] and M Huang[1]

[1] Nanjing Electronic Devices Institute, No.524, zhongshan road,Nanjing, Jiangsu Province, China

zhengfei720@126.com

Abstract. To cope with the special requirements of RF microsystem for ultra-thick wiring layer and large-size line width of microwave signal, based on the damascene process in CMOS manufacturing, the dishing phenomenon and causes of thick copper CMP dishing were discussed. Through using the radio frequency module wiring layout, dishing generated by different line width in copper CMP was studied and compared. The influence of the thickness of electroplated copper on dishing is discussed, and a method to reduce dishing is proposed. In addition, in order to reduce the dishing generated by the Ta-CMP process, the removal rates of Ta, Cu and SiO_2 were tested, and a polishing selection ratio close to 1:1:1 were obtained. Dishing was reduced to 1000 Å. Finally, a three-layer thick copper wiring consisting of a thick dielectric film deposition and etching, thick copper plating and CMP was successfully developed with a total thickness of 10μm.

1. Introduction

The chemical mechanical polishing (CMP) process is one of the most effective global planarization technique [1-2], and CMP is one of the key technology for IC manufacturing. The CMP process also plays a vital role in RF microsystem, 3D stacking, wafer level packaging, TSV interconnect, integrated passive device, multilayer interconnect, and sacrificial layer technology.

Copper CMP is one of the key processes in the damascene process [3-5], because the surface morphology after dielectric deposition, etching and metallization is very poor. The large height fluctuation not only affects the second dielectric deposition, but also seriously affects the integrity of the graphic after lithography. Therefore, global planarization is achieved by copper CMP, which places the metal at a relative level to the surface of the dielectric. During the copper CMP process, both dishing and erosion occurs, which can seriously affect the dielectric deposition and metallization. The copper dishing is influenced by many factors, such as the thickness of the metal layer, the duty cycle and the line width. Many researches on reducing dishing had been proposed [6-7]. But in RF microsystem applications, it is not only necessary to meet the requirements of ultra-thick metal layers for microwave signals, but also to meet the requirements of line width for different signal lines. Therefore, thick metal multilayer wiring based on RF microsystems puts higher requirements on copper CMP.

This study is based on the damascene process in CMOS manufacturing, focusing on the process of thick copper CMP dishing in thick copper wiring process, and discuss the main reasons for dishing. In addition, for the different requirements of the thickness and width of the microwave signal line in the

IOP Conf. Series: Journal of Physics: Conf. Series **1209** (2019) 012010 doi:10.1088/1742-6596/1209/1/012010

RF micro-system, the copper CMP dishing of different line width pattern and plating thickness were discussed. Then a practical method to reduce dishing was proposed. Finally, a three-layer thick copper wiring consisting of a thick dielectric film deposition and etching, thick copper plating and CMP was developed.

2. Theories and Methods

2.1. Damascene process

The dual damascene process is the basis of the three-dimensional integrated wiring process. While forming the planar interconnect, the functions of connecting device or interconnect under the dielectric layer to the surface of the dielectric layer can be realized. Therefore, the copper interconnect is currently using dual damascene process, as shown in Figure 1. A thin layer of silicon carbide is deposited by PECVD as a diffusion barrier and an etch stop layer. Then a certain thickness of silicon dioxide is deposited on the silicon carbide. Etching the vertical micro-via which will be used to connect the multilayer metal interconnection after photolithography, the etching depth of micro-via cannot bigger than the thickness of silicon dioxide layer. Then the trench of planar interconnect is formed by photolithography. The trench of the planar interconnect and the micro-via is realized by etching simultaneously. All the dielectric layers at the micro via position are removed. The diffusion barrier layer, the adhesion layer and the copper seed layer are sputtered by PVD, and then the electroplated copper layer. Finally the copper layer on the oxide surface is removed by CMP, which can realize a planar interconnection of the lateral wiring connecting and a vertical interconnection of the inter-layer metal.

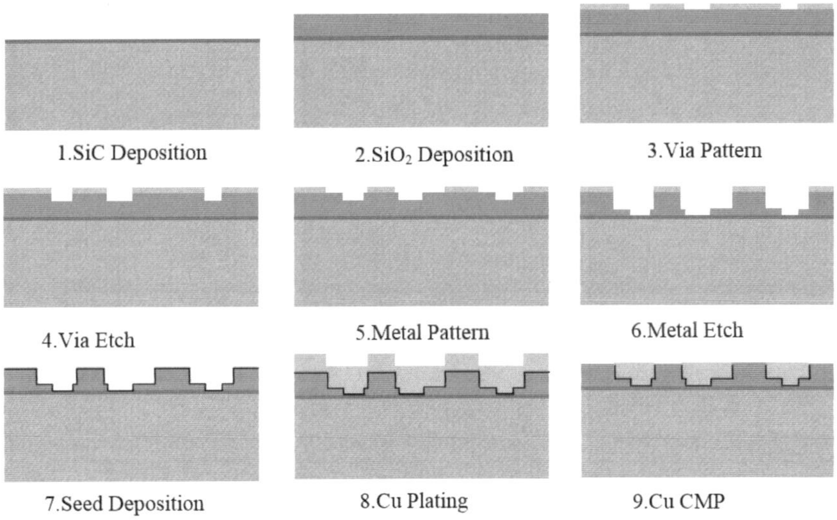

Figure 1. Dual damascene process.

2.2. CMP process

The basic principle schematic of CMP is shown in Figure 2. Under the high-speed rotation of the polishing head and the polishing pad, the polishing liquid and the material to be polished undergo a series of complicated chemical reactions to produce an easily removable substance, which is removed by the abrasive in slurry. Then the newly exposed surface of the polishing material continues to react and be removed. Finally the globe planarization is achieved. Currently, CMP is widely used for planarization of various materials, including polymers, metals, dielectrics, and ceramics, for its excellent performance. The application of CMP can be divided into metal CMP and non-metal CMP, the most representative are silicon dioxide CMP and copper CMP, which represent almost all CMP process principles and methods.

The CMP process involves very complex chemical and mechanical effects. The mathematical theory model is continuously improved on the basis of the Preston model and has not yet been fully understood. The copper CMP process is divided into three steps: 1) polishing copper with high speed for eliminating the steps rapidly; 2) polishing copper with low speed and low pressure until copper is completely removed, for reduction the dishing; 3) barrier Ta is completely removed through specific slurry and process parameters for reduce Dishing. Due to the different physical and chemical properties of different materials, the polishing rates of different materials are also inconsistent. The copper removal rate is too fast, but the barrier layer material and the dielectric layer material are removed very slowly, which is likely to cause local over-polishing. Over polishing will bring in serious dishing and erosion problems, as shown in Figure 3. In addition, the generation of the dishing is also related many factors, such as the deformation of the polishing pad, the type of the polishing liquid, the over-polishing time, the polishing pressure, the relative velocity between the polishing pad and the wafer, the distribution of the polishing particles on the polishing pad, the oxidation in the polishing liquid, the concentration of the material, the pattern density, the copper line width and the plating process of the copper. In this study, we used fixed polishing liquid, polishing pad and disk. Therefore, the method of reducing the dishing is started from the selection ratio of CMP of different materials, the line width of the pattern, and the thickness of the pattern.

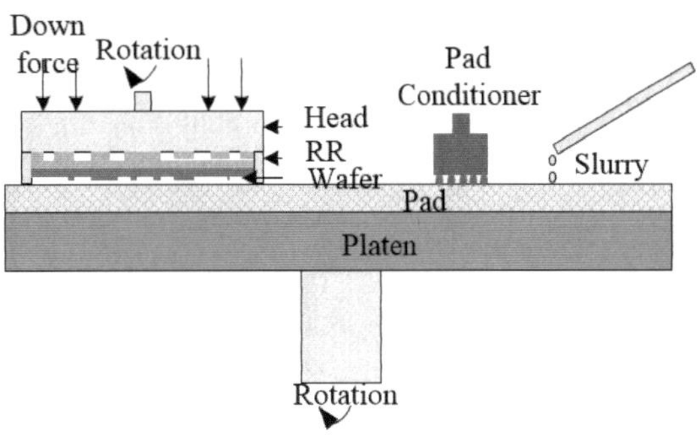

Figure 2. Fundamental schematic of CMP.

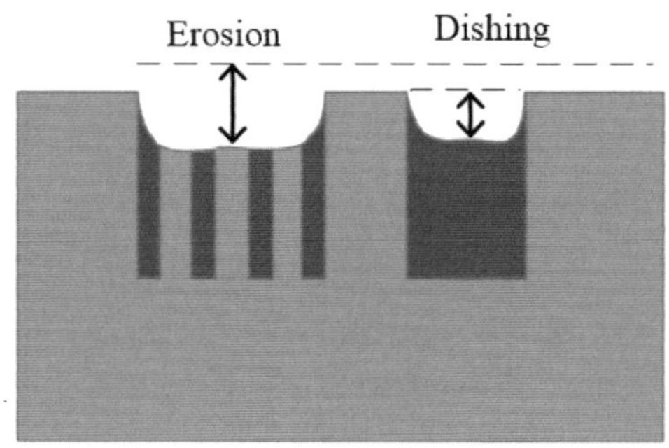

Figure 3. Dishing and Erosion.

3. Experiment and Results

3.1. Experiment condition

All of the CMP experiments were complemented by AMAT Mirra Mesa. The equipment has three polishing platens. Platen1 and platen2 are for Cu polishing, platen3 is for barrier metal and oxide polishing. In order to achieve controllable non-uniformity, 5 zone polishing head is used. And we use IC1000 pad, 3M 165A disk, and commercialization slurry. We use the optical microscope and scanning electronic microscope (SEM) to characterize the copper surface topography after CMP. Platen1 and Platen2 used the same commercial polishing slurry, and the proportion of the slurry was the same. So, the polishing rate and polishing non-uniformity of copper CMP were adjusted by the down force, the platen rotation rate, the head rotation rate and the flow rate.

Polishing pressure is a very important parameter in the chemical mechanical polishing process, which directly determines the polishing rate and polishing effect. Down force mainly determines the polishing rate and the non-uniformity of CMP. The removal rate varies with down force under different slurry flow rate as shown in Figure 4. When rotation rate of the platen and head is fixed, experiments of copper CMP have shown that the removal rate increases as increasing the down force, but an increase in the removal rate causes scratches and bigger dishing. The removal rate increases as increasing the flow rate of the slurry, but the removal rate does not increase if the flow rate increases to a certain extent, because the chemical reaction is already saturated. It can be seen from the figure that the removal rate can be maximized and the polishing effect is best when the flow rate is 200 ml/min.

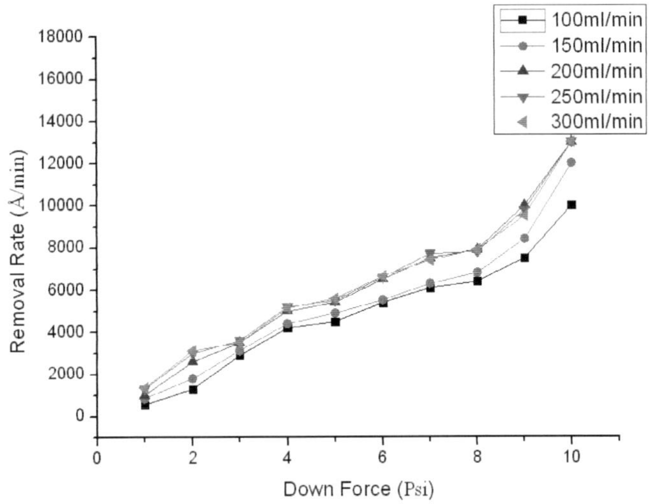

Figure 4. Removal Rate versus down force.

3.2. The effect of line width on dishing

In order to meet the special requirements of the RF microsystem for the high power and microwave performance of the interconnect, the three-layer wiring has a line width ranging from 2μm to 600μm and a line spacing ranging from 2μm to 500μm, which includes strip, circle and square structure, as shown in Figure 5. Therefore, further process optimization is performed for the influence of the line width on dishing.

Copper CMP experiments with a large number of different line widths show that the generation of dishing is the result of the combination of down force and line width, as shown in Figure 6. Dishing increases as increasing of the line width. When the line width increases to more than 300um, the effect of the increase in line width on dishing is relatively reduced. In addition, dishing increases significantly as increasing of down force. We selected several down force to CMP. When the down force is less than 3Psi, the dishing is greatly reduced. Because the removal rate is very low when the down force is too small, the dishing decreasing by reduction the down force is not obvious. Therefore, if the line width cannot be changed, lowering down force is the best way to reduce dishing. We can use low down force on platen2 for much more less dishing.

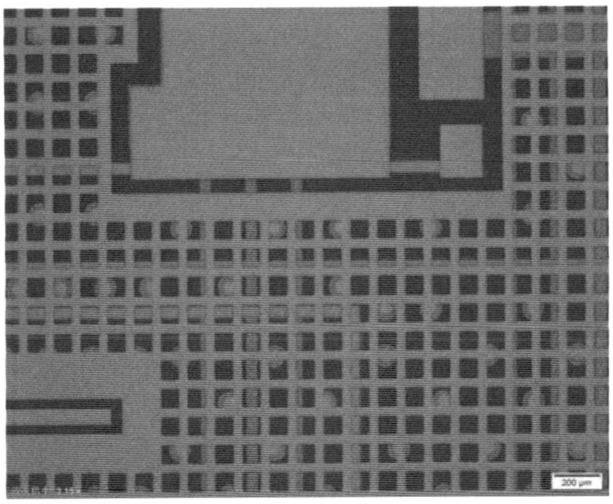

Figure 5. Pattern of three-layer inter-connection.

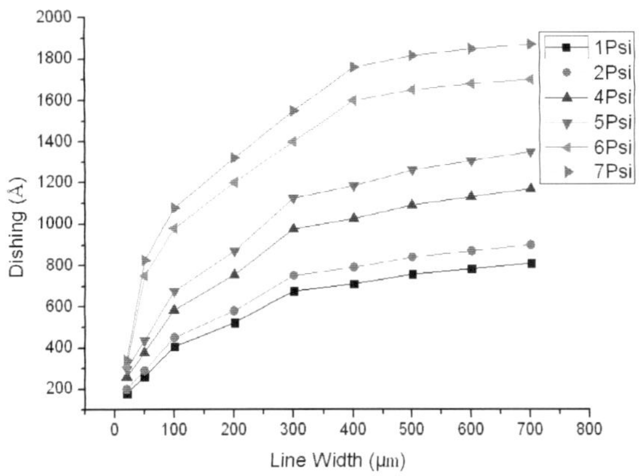

Figure 6. Dishing versus Line Width.

3.3. The effect of plating thickness on dishing

In the RF microsystem, in order to reduce the loss of microwave signal transmission, the thickness of the metal layer is generally greater than 2μm, or even more than 10μm. Thick metal means that there are deeper steps after plating, which places higher demands on copper CMP, as shown in step D in Figure 7. As the process progresses during the copper CMP process, the step of the plating trench will gradually become smaller until the step is removed completely. The thickness of the electroplated Cu affects the effect of the grooving, which in turn affects the CMP process. For a 2μm wiring thickness, the effect of plating thickness on dishing is shown in Figure 8. The patterns with line widths of 2μm, 10μm, 50μm, and 100μm were selected. It can be seen that when the thickness of the electroplated Cu is 2 to 3 times the thickness of the copper wiring, the reduction of dishing is the largest and the value of dishing is the smallest.

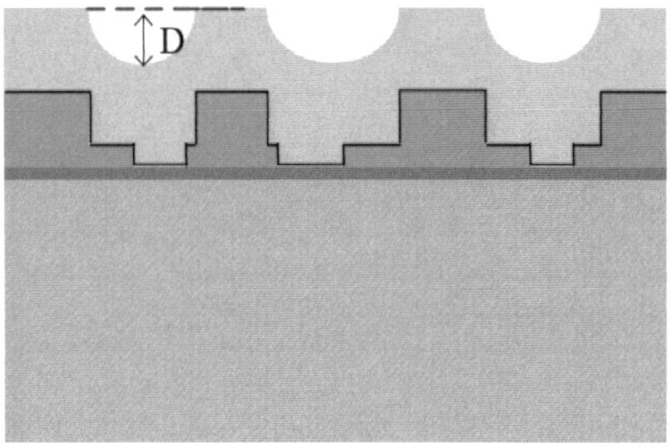

Figure 7. Step of Plating.

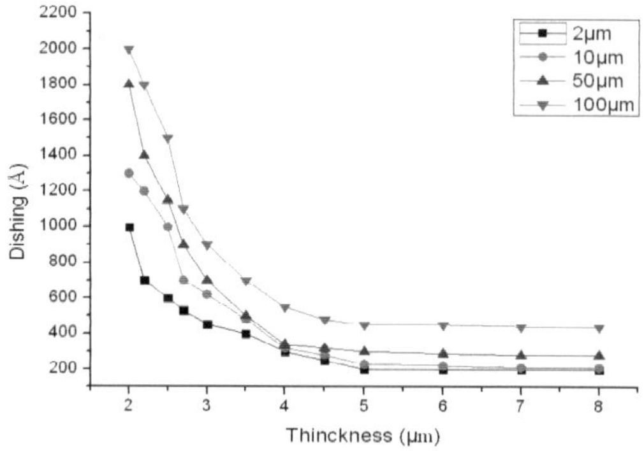

Figure 8. Dishing versus plating thickness.

3.4. Removal rate of Ta-CMP

In order to minimize the influence of dishing increase due to different removal rates of Cu, Ta and SiO2, the removal rate of three materials was studied on Platen3. After optimization of the process parameters, the polishing selection ratio of the three materials is close to 1. As shown in Figure 9, the removal rates of Cu, Ta, and SiO_2 on the 8-inch silicon wafer are all close to 500 Å/min, and the Ta removal rate has a good non-uniformity, NU < 2%. Through the optimization of process parameters, dishing can be reduced to less than 1000Å.

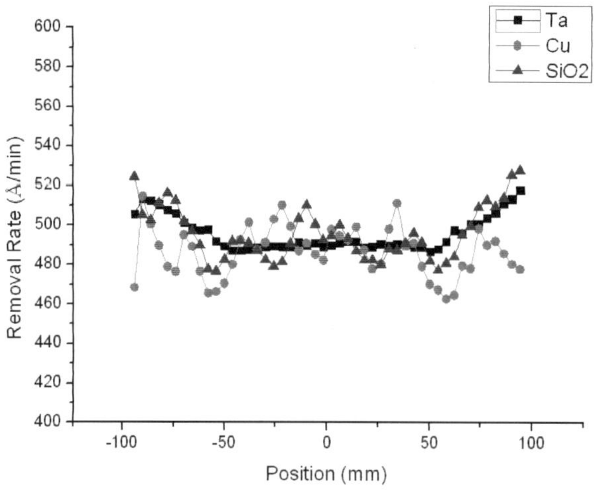

Figure 9. Removal Rate on platen 3.

3.5. Three-layer wiring

Through the optimization of the above CMP process parameters, the three-layer interconnect wiring process is completed based on the module layout of the T/R module. Figure 10 is the profile of the three-layer interconnection. The process includes three layers of wiring of Metal1, Via2, Metal2, Via3, and Metal3. It was completed by 3 times of dielectric deposition, metallization, and CMP, and the final thickness of the wiring was 10μm. The thickness of Metal1, Via2, and Metal2 is 2μm, the thickness of Metal3 is 3μm, and the thickness of Via3 is 1μm. The SEM image of the three-layer wiring is shown in Figure 11. Because of the variation of deposition of and measurement, the thickness is very close to the design value. The thickness of Metal1 is 1.99μm, the thickness of Via2 is 2.00μm, the thickness of Metal2 and Via3 is 3.00μm, and the thickness of Metal3 is 3.00μm.

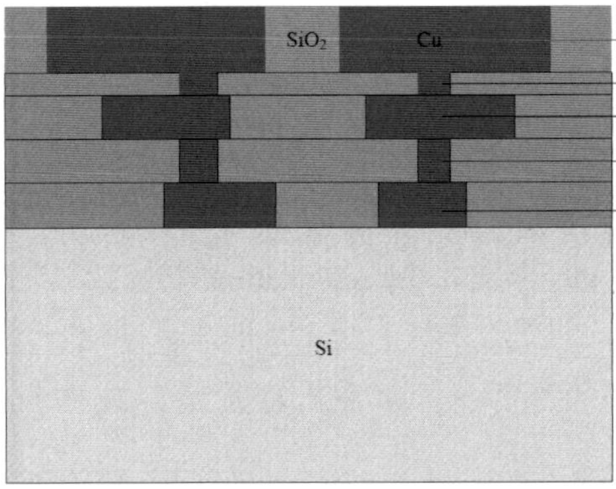

Figure 10. Profile of the three-layer interconnection.

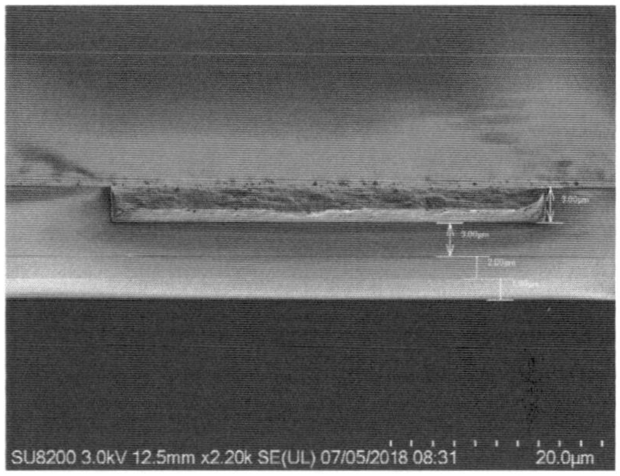

Figure 11. SEM image of the three-layer interconnection.

Through the further optimized CMP process parameters and plating conditions, the three-layer wiring is completed. All the lines are complete in structure, the lines are clear, the surface of the Cu line has no defects, and the electrical performance can meet design requirements. After measurement, the Cu line dishing is less than 1000 Å. The SEM image of the cross section is shown in Figure 12. It can be seen that the three-layer structure is complete and the micro-via has no fracture.

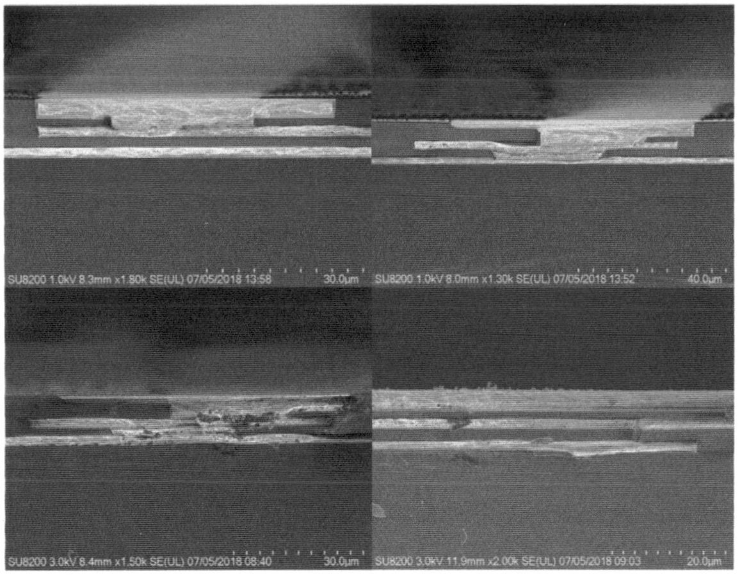

Figure 12. SEM image of three-layer interconnection profile.

4. Conclusion

IOP Conf. Series: Journal of Physics: Conf. Series **1209** (2019) 012010 doi:10.1088/1742-6596/1209/1/012010

In this paper, we analyze the dishing problems in thick copper CMP process. By optimizing the process parameters, setting the low down force, selecting 2 to 3 times the thickness of the electroplated copper, the dishing effect of the large-size line width and the thick copper CMP can be greatly reduced. And through the optimized copper CMP process, a three-layer wiring structure based on the RF module is realized.

Acknowledgments
The authors wishing to acknowledge assistance or encouragement from colleagues.

References

[1] Zhang Kailiang, Liu Yuling, Wang Fang, Chemic-Mechanical Polishing of Silicon Wafer in ULSI, *Chinese Journal of Semiconductors*, 2004, 25 (1):117

[2] Ng D, Kulkarni M, Johnson J, Oxidation and removal mechanisms during chemical mechanical planarization, *Wear*.2007, 263: 1477-1483

[3] Levrecht von Trotha, George Moersch, Gerfried Zwicker, Advanced MEMS fabrication using CMP, *Semiconductor International*, (81112004)

[4] J. Knickerbocker, P. Andry, B. Dang, R. Horton, et all, 3D Silicon Integration, Proc. of 58th *Electronic Components and Technology Conference*, LakeBuena Vista, FL, May 2008, p.538-543

[5] Beomyoung Park, et all, A Study on the Reduction of Dishing and Erosion Defects in Tungsten CMP, *J. of KSPE*

[6] Yongjin Guo, Abhijit Chandra, Ashraf-F. Bastawrosb, Analytical Dishing and Step Height Reduction Model for CMP With a Viscoelastic Pad, *Journal of The Electrochemical Society*, 151 (9) G583-G589 (2004)

[7] Runzi Chang, C.J Spanos, Dishing-Radius Model of Copper CMP Dishing Effects, *IEEE Transactions on semiconductor manufacturing*, VOL.18, NO.2, MAY 2005

Optimization of fluorescent aptamer sensor for ATP detection

L j Ren [1], X Wei [1], X x Hang [1], P Zhang [1], J t Zhang [1], Q f Zhang [1], L y Jiang[1*]
1 Department of Electric Information Engineering, Zhengzhou University of Light Industry, zhengzhou, Henan province, China

E-mail: jiangliying@zzuli.edu.cn

Abstract. Abnormal concentrations of ATP are associated with many diseases and cancers, and quantitative detection of ATP is thus of great importance for disease diagnosis and prognosis. In order to obtain a simple and quickly detection method of ATP by fluorescence aptasensor, the principle of ATP detection based on resonance energy transfer between FAM and GO was studied. And then the optimization of process conditions was carefully researched. The results showed that when the concentration and reaction time of GO was 10μ g/mL and 5min, respectively, the quenching light intensity was the easiest to recover. With the optimized parameters, the concentration of ATP in the range of 1-250μ mol/L had a good linear relationship with the fluorescence intensity, and the detection limit was 1μ mol/L. Besides, the selectivity experiments proved the selectivity of the constructed aptasensor toward ATP. This method creates the opportunity to detect a broad range of targets by using appropriate aptamers, holding great potential in the field of food safety and biomedical diagnostics.

1. Introduction

Abnormal concentrations of ATP are associated with many diseases and cancers, such as Hypoxia, hypoglycemia, Parkinson's disease and some malignant tumors [1-4]. In addition, the utilization efficiency of ATP is also used as an important indicator of cell survival energy and cell injury [5]. So quantitative detection of ATP is thus of great importance for disease diagnosis and prognosis [6]. At present, conventional methods, such as chemiluminescence [7-9], high performance liquid chromatography (HPLC) [10-12], and mass spectrometry [13,14] have been widely used in the detection of ATP. However, these methods are not suitable for real-time analysis because of their costly apparatus and complicated pretreatment procedures. So there is a demand for biosensors capable of rapid, accurate and reliable determination of low concentrations of ATP.

Aptamers have prominent advantages in detecting small molecular biological targets due to their high stability, low cost, simply synthesis, easy marking, no immunity source and toxicity [15-17]. But, aptamers cannot produce any light or electrical signals. Fluorescence aptasensor has been widely researched for biomolecular detection in in clinical diagnostics and food safety fields, because of its high sensitivity from fluorescence detection and its high selectivity according to nucleic acid aptamer [18]. Recently, fluorescent aptasensors have been widely researched for detections of many biomolecules, such as AFB1 [19,20], DA [21-23], AD [24] and so on. These aptasensors recognize and capture targets with aptamers which exhibit their high affinity and specificity, and then measuring the concentrations of target molecules by fluorescent signal variation.

In the present work, we report a simple and quickly detection method of ATP by fluorescence aptasensor based on resonance energy transfer. And by optimizing the detection parameters, a wide detection range with high detection limit of ATP were obtained. The principle of ATP detection was shown in Figure 1. When the graphite oxide is added to nucleic acid aptamer solution modified by FAM, luorescence intensity of FAM will decreases significantly due to resonance energy transfer.After the target is added, the aptamer will be first combined with the target, leading to FAM dissociation from graphene oxide and fluorescence recovery. The amount of restoring fluorescence intensity is related to the concentration of the target.

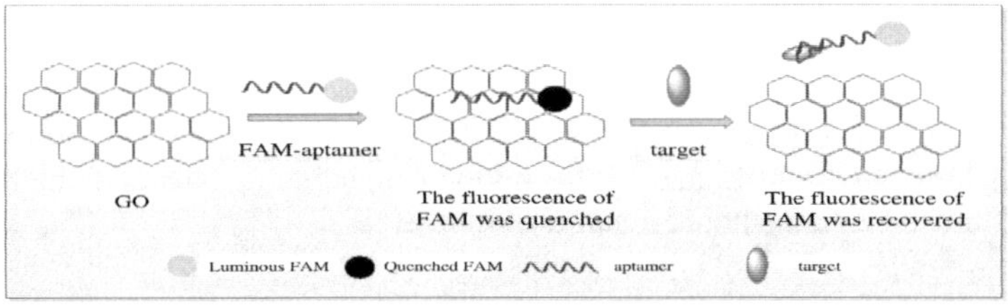

Figure 1. Schematic diagram of ATP detection principle.

2. Materials and Methods

2.1. Reagents and materials

Tris (2-carboxyethyl) phosphine (TCEP) and adenosine triphosphate (ATP) aptamer were purchased from Shanghai Sangon Biological Engineering Technology & Services Co. Ltd. (Shanghai, China). ATP aptamer: 5'-FAM-ACCTGGGGGAGTATTGCGGAGGAAGGT-3'. ATP were purchased from Beijing Solarbio Science & Technology Co., Ltd (Beijing, China). Phosphate buffer solution (PBS,10mM, pH 7.4), Tris-acetate buffer (10mM, pH 5.2). All chemical reagents were of analytical grade and the ultrapure water used throughout the study was prepared by PURELAB Option-R (ELGA LabWater, UK).

2.2. Instruments

Fluorescence measurements were performed on a F-7000 fluorescence spectrophotometer (Hitachi, Japan). Eppendorf centrifuge 5418(Hamburg, Germany) was used for the centrifugation. All pH values were acquired on a FE-20K pH-meter (METTLER TOLEDO, Switzerland). GL-16II centrifuge (Shanghai Anting Scientific Instrument Factory) and 07HWS-2 digital thermostat magnetic stirrer (Hangzhou Instrument Motor Co., Ltd.) were also used in the experiment.

2.3. Preparation of detection probes

Briefly, the aptamer stored at -20 °C was centrifuged for 30 s (12,000 rpm), then added Tris-HCl to bring the final concentration of FAM-aptamer to 1 μmol/L. The solution was incubated at room temperature for 2 hours under dark. The FAM-aptamer were mixed with the oxidized octaene solution, stand at room temperature for a period of time to immobilize the FAM-aptamer on the graphene oxide and the fluorescence of the FAM was quenched (Figure 1).

2.4. Fluorescent detection of ATP

Various concentrations of ATP (1 to 250μmol/L) were prepared in a stock solution (50mmol Tris, 30mmol NaCl, 50mmol KCl) and added to the prepared probe solution. After incubation for a period of time, 400μL was taken into a fluorescence spectrometer to detect fluorescence. Then, the ATP working solution was used as a blank control group to calculate the fluorescence intensity of the different concentrations of the test solution and draw a standard curve.

3. Results and Discussion

3.1. Feasibility of the fluorescent aptasensor

IOP Conf. Series: Journal of Physics: Conf. Series **1209** (2019) 012011 doi:10.1088/1742-6596/1209/1/012011

The fluorescence emission spectra of FAM-DNA, FAM-DNA+GO, FAM-DNA+GO+ATP were shown in Fig.2. The fluorescence intensity of FAM sharp droped with the addition of GO (10μg/mL,900μL), which due to the fluorescence resonance energy transfer from FAM to GO. When ATP (1μM, 100μL) was added, the ATP was specific binding with its aptamer. Thus, the FAM-DNA would detach from the surface of GO, and the fluorescence intensity would recovered.The amount of restoring fluorescence intensity would be related to the concentration of the ATP.

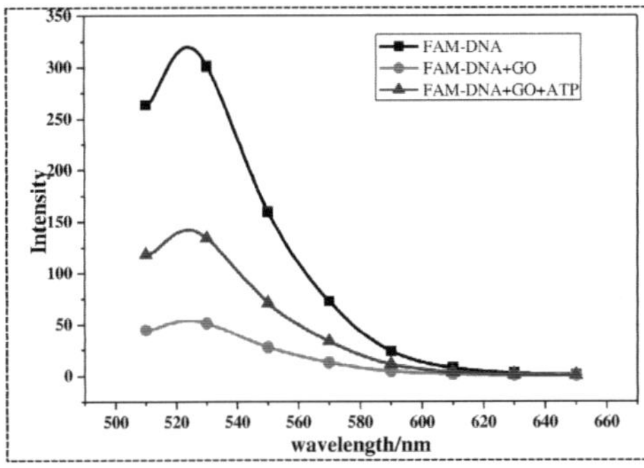

Figure 2. The fluorescence emission spectra of FAM-DNA, FAM-DNA+GO, FAM-DNA+GO+ATP.

3.2. Optimization of experimental conditions

In this method, The degree of fluorescence quenching will directly influence the intensity of light recovery, which will affect the measurement results. The degree of fluorescence quenching is affected by the concentration and reaction time of GO. To optimize these parameters, fluorescence spectra of ATP aptamer modified by FAM solution added by different concentrations of GO solution (2,5,10,12,15,20μg/mL, Figure 3) and different reaction times (1-15min, Figure 4) were measured. The results showed that when the concentration and reaction time of GO was 10μg/mL and 5min, respectively, the quenching light intensity was the easiest to recover.

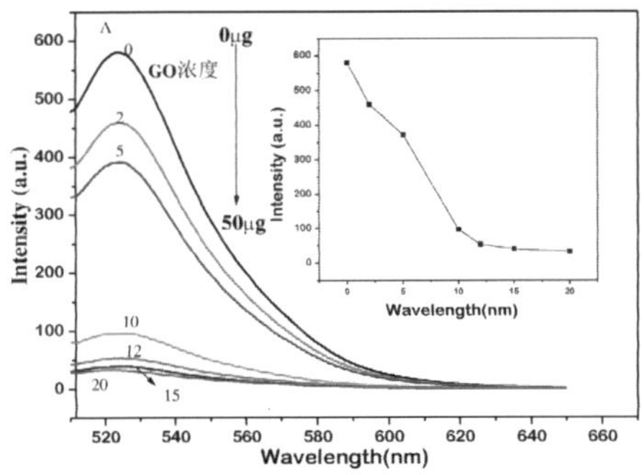

Figure 3. Fluorescence spectra of ATP aptamer modified by FAM solution added bydifferent concentrations of GO solution.

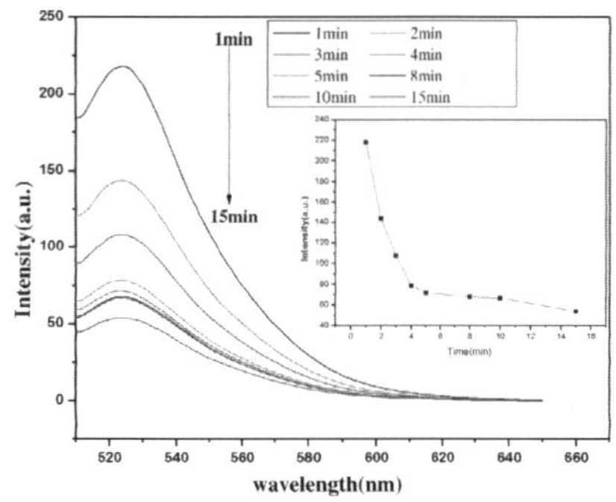

Figure 4. Fluorescence spectra of ATP aptamer modified by FAM solution added by the same concentration of GO solution with different reaction times.

IOP Conf. Series: Journal of Physics: Conf. Series **1209** (2019) 012011 doi:10.1088/1742-6596/1209/1/012011

3.3. Quantitative detection of ATP

With the optimized parameters, differernt concentrations of ATP were measured. Fig. 5(a) showed the fluorescence spectra of restoring fluorescence intensityC QDs/AuNPs-aptamer/DA system with different concentrations (0.05, 25, 100, 150, 200, 250 μM) of DA The calibration curve of ATP concentration and fluorescence intensity was shown in Figure 5(b). The results showed that the concentration of ATP in the range of 1-250μmol/L had a good linear relationship with thefluorescence intensity. The linear regression equation was Y=1.5266X+96.16572 (Y was the fluorescence intensity and X was the ATP concentration). The regression coefficient is 0.94905, and the detection limit (The minimum that the sample can be accurately detected) was 1μmol/L.

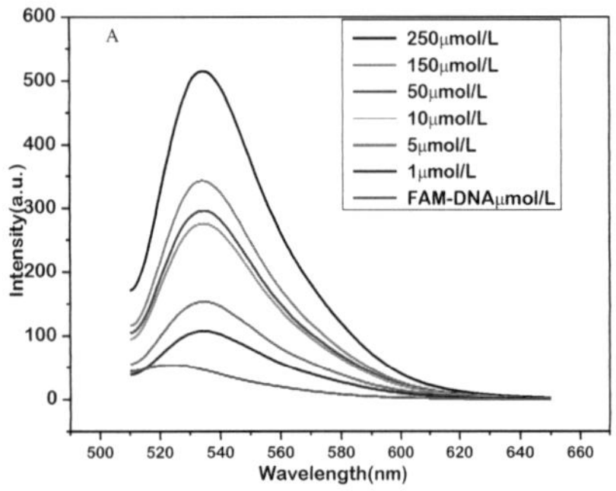

Figure 5 (a). Fluorescence signal curves of ATP.

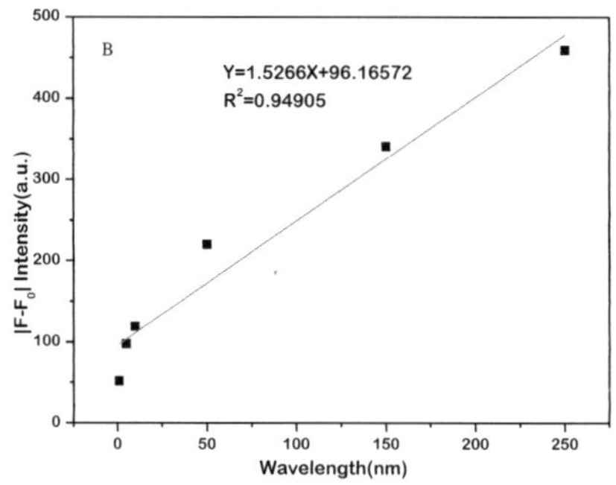

Figure 5 (b). Calibration curve of ATP concentration and fluorescence intensity.

.

3.4. Selectivity of DA based on the assay

In order to analyze the specificity of this method to ATP, Na+, K+, DA ,Biotin, adrenaline (AA), and ATP were all tested with the same parameters (100μL, 1μM) and the same experimental conditions. Compared with other subjects, ATP has significant specificity which were shown in Figure 6.

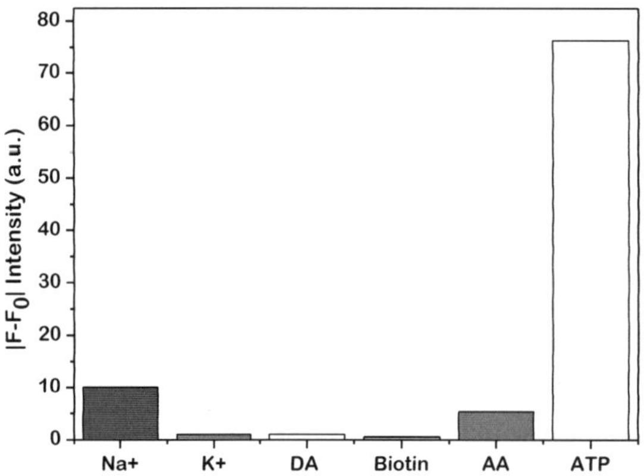

Figure 6. The specific detection of ATP (1μmol/L).

4. Conclusions

In conclusion, we report a simple and quickly detection method of ATP by fluorescence aptasensor based on resonance energy transfer. The principle of ATP detection was studied, the detection parameters were optimized, and then different concentrations of ATP were detected. Finally, The specificity of this method for ATP detection was tested. The results showed that with the optimized parameters, the concentration of ATP in the range of 1-250μmol/L had a good linear relationship with the fluorescence intensity, and the detection limit was 1μmol/L. Besides, compared with other subjects, ATP has significant specificity of this method, which may be applied for detection of pathogenic bacterium in food safety and biomedical diagnostics.

Acknowledgments

The work for this paper was supported by the National Natural Science Foundation of China (Grant Nos. 61801436), Science and Technology Innovation Outstanding Talents Project of Henan Province (Grant No. 184200510015), Henan Province Science and Technology Plan Project - International Cooperation Project (Grant No. 182102410093), Key research projects in colleges and universities of Henan Province (Grant No. 18A510021), Doctoral Research Fund Project of Zhengzhou University of Light Industry (Grant No. 2015BSJJ023, Grant No. 2014BSJJ041), Henan Key Laboratory of Biomolecular Recognition and Sensing (Shangqiu Normal University) and Graduate Technology Innovation Fund Project of Zhengzhou University of Light Industry (Grant No.2018022)

References

[1] T Y Aw, D P Jones 1985 *Am. J. Physiol.* 249 C385.
[2] K T Bush, S H Keller and S K Nigam 2000 *J. Clin. Invest* **106** 621-6.
[3] C. Ma, X. Yang, K. Wang, Z. Tang, W. Li, W. Tan and X. Lv 2008 *Anal. Biochem.* **372** 131-3.
[4] Ferrer, E. Perez, E. Dalfó and M. Barrachina 2007 *Neurosci. Lett.* **415** 205-9.
[5] Cruz-Aguado, J. A., Chen, Y., Zhang, Z., Elowe, N. H., Brook, M. A. and Brennan, J. D 2004 *J. Am. Chem. Soc* **126** 68-78.
[6] Li. F., Hu X., Wang F., Zheng B., Du J and Xiao D 2018 *Talanta.* **179** 285.
[7] Fan G. C., Zhao M., Zhu H., Shi J. J., Zhang J. R and Zhu, J. J 2015 *J. Phys. Chem. C.* **120** 29.
[8] Motoyoshiya J, Wada J Y., Itoh K., Wakabayashi K., Maruyama T and Ono K 2018 *S.A.P:M&B.S.* **195** 223.
[9] Sun Y, Wang Y, Li J, Ding C, Lin Y and Sun W 2017 *New J. Chem* **10.**1039.
[10] F Ozogul, T Kda, P.C. Quantick and Y Ozogul 2001 *Int. J. Food Sci. Technol.* **35** 549–54.
[11] Liu S, J Xu, J J, Ma C. L and Guo C F 2018 *Food. Chem.* **10.**1016.
[12] Onozato M, Tanaka Y, Arita M, Sakamoto T, Ichiba H and Sadamoto K.2018 *Pedlatr. Nephrol* **12** e00099.
[13] Castro-Puyana M, Pérez-Míguez R, Montero L and Herrero M 2017 *Trac.Trend. Anal. Chem* 93.
[14] Zwickel T, Klaffke H, Richards K and Rychlik M 2016 *J. Chromatogr. A.* **1455** 74-85.
[15] Wang RE, Zhang Y, Cai J, Cai W and Gao T 2011 *Curr. Med. Chem* **18** 4175-84.
[16] Citartan M, Gopinath SC and Tominaga J 2012 *Biosens. Bioelectron* **34** 1-11.
[17] Liu X, Li W, Xu X, Zhou J and Nie Z 2012 *M ethods Mol. Biol.* **800** 119-32
[18] Zhao Q, Lv Q and Wang 2015 *Biosens. Bioelectron* **70** 188-93.
[19] Alzubaidi, Labeeb A, M. T. Almossawei and S. Y. Abduljaleel 2017 *Antibiotiki* **4.4** 33-7.
[20] Xu W, Xiong Y, Lai W, Xu Y, Li C and Xie M 2014 *Biosens. Bioelectron.* **56** 144-50.
[21] Zhao J, Zhao L, Lan C and Zhao S 2016 *Sensor. Actuat. B. Chem* **223** 246-51.
[22] Govindaraju S., Ankireddy S R, Viswanath B, Kim J and Yun, K 2017 *Sci. Rep-Uk* **7** 40298.
[23] Zhao L, Xin X, Ding P, Song A, Xie Z and Shen J 2016 *Anal. Chim. Acta* **926** 99-106.
[24] Wang M, Li Y, Wang L and Su X 2017 *Anal. Methods.* **10** 1039.

IOP Conf. Series: Journal of Physics: Conf. Series **1209** (2019) 012012 doi:10.1088/1742-6596/1209/1/012012

Design and optimization of a novel high isolation low insertion loss RF MEMS single-pole double-throw switch

Zh Xiao [1], Zh l Deng [2] and Y ch Wang [3]

[1,2,3] School of Electronic Engineering, Beijing University of Posts and Telecommunications, Beijing 100876,China;

E-mall: 281590917@qq.com; Tel.: +86-150-3198-0711

Abstract. This paper designs an RF MEMS single-pole single-throw and single-pole double-throw switch based on the Ka-band on the CPW. The electrodes of the switch are sandwiched by two dielectric layers. Each SPDT switch has a resonant frequency of approximately 35 GHz and a bandwidth of 20 GHz with an isolation greater than -38 dB and an insertion loss less than -0.8 dB. By changing the state of the SPDT switch, one is in the up state and the other is in the down state, it can achieve the function of switching on and off. The SPDT switch also operates at around 35 GHz with a return loss greater than -23 dB, an insertion loss of less than -1 dB, and an isolation greater than -38 dB.

1. Introduction

Compared with PIN diodes and FETs, RF MEMS switches have smaller size, lighter weight, are not sensitive to acceleration, have no DC loss at microwave frequencies, and have high isolation and low insertion loss [1], which is communication. The system is widely used in phased array radar [2], electronic warfare, satellite communications [3] and other important control units. Compared with the microstrip line, the CPW transmission line has the signal line and the ground line on the same side of the substrate. When the series and parallel circuit components are installed, the substrate is not required to be perforated [4]. These features are helpful for the wide application of CPW transmission line in the field of RF MEMS. Compared with other processing technologies, silicon-based MEMS processing technology is very mature. Therefore, the RF MEMS switch of silicon-based CPW transmission line has been a hot spot in international scientific research. For capacitive switches, the higher the capacitance ratio, the better the RF performance of the switch will be. The use of large beam displacements [5] and relatively thin dielectric layers can effectively increase the upper and lower capacitance ratios. However, the driving voltage of the MEMS switch increases significantly as the beam height increases; Secondly, a thin dielectric layer can cause the switch to break down and fail. High dielectric constant materials can increase the on-state capacitance of the switch and increase the insertion loss of the switching operating band. Therefore, the low-loss, high-isolation large capacitance is urgently needed to solve the key technology of the structural design of the RF MEMS switch. In the paper, Koul S K designed a high capacitance ratio parallel MEMS switch based on CMOS technology, which has a return loss of 12.7 dB~17 dB from 10 GHz to 20 GHz. The Ku-band (13 - 17 GHz) SPDT switch [6] for the T-module application proposed by Sukomal Dey et al. used a series-tap mode of operation; Deepak Bansal et al. proposed a Y-type anchorless slab SPDT switch using a single series capacitor switch mode of operation [7]; Although the above switch design has good switching performance, it is not suitable for the high frequency band, because the touch switch is easy to contact and generate electric spark at a higher

IOP Conf. Series: Journal of Physics: Conf. Series **1209** (2019) 012012 doi:10.1088/1742-6596/1209/1/012012

frequency, thereby shortening the switch life. Professor Gou et al form Harbin Institute of Technology have simulated SPDT switches based on microstrip line series MEMS and CPW parallel MEMS switches [8] and found that the impedance was difficult to match during the design process due to the complexity and asymmetry of the switch structure.

In this paper, some research on RF MEMS switches and CPW transmission lines have been carried out based on some of the above problems, and the details, analysis and simulation of the switch design are introduced. In order to solve the impedance matching problem of the discontinuous structure of Ka-band power distribution network, the transmission line theory is used to design a reconfigurable feeder power network model based on SPDT switch according to theoretical analysis, it is found that the switch has high isolation and low insertion loss in both operating states.

2. Circuit model of MEMS switch

The MEMS shunts switch can be integrated on a coplanar waveguide (CPW) or microstrip line if a 1/4 wavelength stub is to be connected to an anchor on the microstrip line. In this paper, a 50 ohms matched CPW transmission line is used for circuit model analysis. The center bandwidth of the CPW is W, the gap width is G, and the thickness is 1μm, and 400μm is used as the thickness of the silicon substrate. In the CPW transmission line, the thickness of the dielectric layer is t_d, and the dielectric constant is ε_r, where g is the distance of the switch beam from the dielectric layer. The beam has a length L, a width w, and a thickness t. These are some of the basic dimensions of RF MEMS switches. The side view of the MEMS capacitor parallel switch is shown in Figure 1.

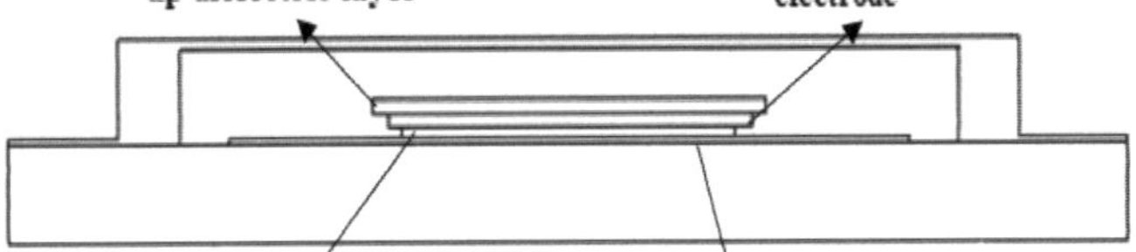

Figure 1. Side view of the MEMS parallel variable container.

Typically, the MEMS switch beam is connected between the transmission line and ground, and the switch anchor area is connected to the CPW ground. When a DC voltage is applied between the MEMS beam and the CPW transmission line, the resulting static force causes the beam to bend downward, leaving the switch in the down state. When the DC bias voltage is not applied, the light beam returns to the initial state (down state) by the elastic restoring force.

The upper capacitor C_u of the switch is composed of a parallel plate capacitor C_p and an edge capacitor C_e. The edge capacitance is about 20% to 40% of the parallel plate capacitance, and its value varies with the height of the beam. In this paper, the structure of the double-layer dielectric clamped electrode is used as the dielectric layer. So the parallel plate capacitor C_p is equivalent to two capacitors in series. The parallel plate capacitance is given by the following equation (1), and the capacitance ratio is given by (2), where A_{up} and A_{dn} is the opposite area of parallel plates, and t_d is the dielectric layer thickness.

$$C_{dn} = \varepsilon_0(A_{dn}[\tfrac{t_d}{\varepsilon_r}]^{-1})/2 \quad (1)$$

$$C_{up} = \varepsilon_0 A_{up}\varepsilon_r[g_0\varepsilon_r + 2t_d]^{-1} + C_f \quad (2)$$

When the edge effect is neglected, the ratio of the upper state C_{up} to the lower state capacitance C_{dn} of the above varactor can be approximated as:

$$C_{dn}/C_{up} = A_{dn}/A_{up}\left[1 + g_0\frac{\varepsilon_r}{t_d}\right] = 1 + g_0\frac{\varepsilon_r}{2t_d} \quad (3)$$

IOP Conf. Series: Journal of Physics: Conf. Series **1209** (2019) 012012 doi:10.1088/1742-6596/1209/1/012012

The upper state and the lower plate are equal in area, that is, $A_{up}=A_{up}$. According to the structural size of the equivalent CPW, the characteristic impedance can be calculated by the formula (4) [9], and the isolation is obtained by the formula (5).

$$Z_0 = \begin{cases} \frac{60}{\sqrt{\varepsilon_e}} ln\left(\frac{8d}{W} + \frac{W}{4d}\right) & (\frac{W}{d} \leq 2) \\ \frac{120\pi}{\sqrt{\varepsilon_e}\left[\frac{W}{d}+1.393+0.667\, ln\left(\frac{W}{d}+1.444\right)\right]} & (\frac{W}{d} > 2) \end{cases} \quad (4)$$

$$S_{21} = \frac{1}{1+j\omega C_d Z_0/2} \quad (5)$$

We use ADS software to simulate the circuit model of the MEMS switch in the Ka-band by typical values, then observe the simulation results of the scattering parameters to obtain high isolation and low insertion loss.

The beam elastic modulus of a MEMS beam is given by:

$$k = 4Ew(\frac{t}{l})^3 \quad (6)$$

Figure 2. Vertical view of MEMS beam.

After the elastic coefficient is obtained, the magnitude of the driving voltage can be calculated according to the formula (7). The parameters in equation (7) are given in Table 1.

$$V_P = \sqrt{\frac{8kg_0^3}{27Ww\varepsilon_0}} \quad (7)$$

3. Structure and design of RF MEMS switch

The material of the CPW transmission line and the MEMS switch including the beam and the anchor is known as gold. We use Si3N4 as the dielectric layer. We can deform the switch beam because increasing the complexity of the beam deformation can effectively change the inductance. The resonant frequency of the switch increases as the inductance increases. We can use the above method to adjust the resonant frequency of the switch. All parameters used in the RF MEMS switch model are shown in Table 1. Figure 3 shows the structural model of the switch.

Table 1. Optimization parameters related to the switch

Symbol	Describe	Value
W_b	width of the beam	120μm
L_b	length of the beam	300μm
W_h	width of the releasing hole	15μm
L_h	length of the releasing hole	15μm
W_{d1}	width of the up dielectric layer	130μm
L_{d1}	length of the up dielectric layer	140μm
W_{d2}	width of the down dielectric layer	140μm
L_{d2}	length of the down dielectric layer	150μm
$t_{d,}$	thickness of the dielectric layer	0.2μm
W	width of center transmission line	100μm
G	width of the gap	60μm
L_a	length of the anchor	300μm

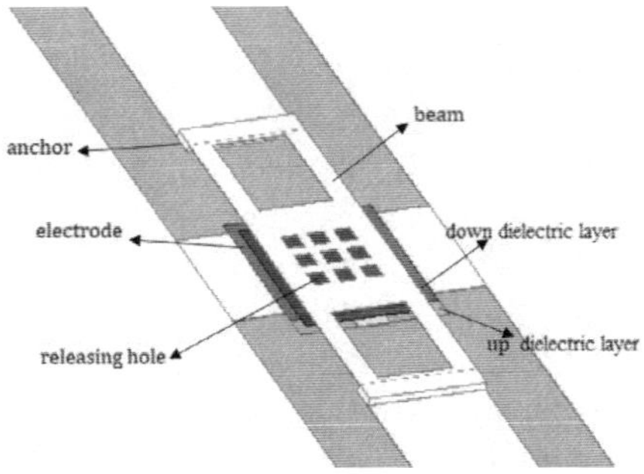

Figure 3. Switch structure modelling.

Observing the simulated S-parameter results, it can be seen from Figure 6 that when the switch is in the upper state, the insertion loss is about -0.1 dB to -0.2 dB in the bandwidth of 30 GHz to 40 GHz; when the switch is in the down state, isolation is approximately -34 dB to -38 dB over the same bandwidth.

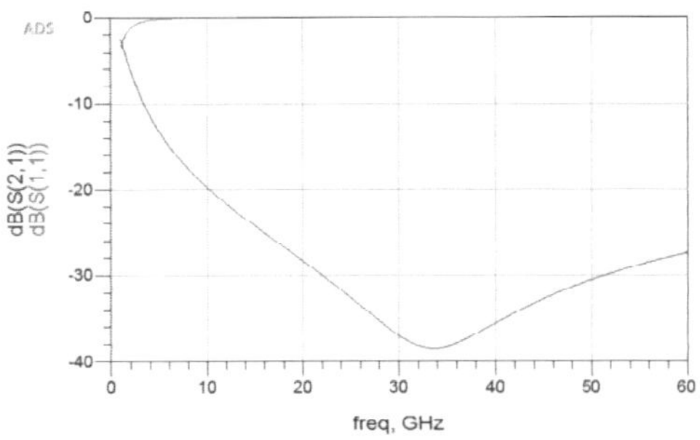

Figure 4. S-parameters graph.

4. Theoretical Analysis and Design of MEMS Single-Pole Double-Throw Switch

In this paper, the T-shaped CPW transmission line is used to design the SPDT switch. The two incident waves can be obtained by equation(8) through a transmission line with a characteristic impedance of 70.7ohms to make the output wave in impedance matching. From this we can get the width W1 of the segment of the transmission line. We place the two switches at a symmetrical position of 1/4 wavelength from the point of convergence. When the switch is in the down state, it is actually equivalent to an open circuit at the point of convergence, as shown in equation(9). Therefore, theoretically, there is no energy loss. Wherein $Z_l = 0$, $l = \lambda/4$, therefore $Z_{in} = \infty$ which is close to open circuit.

$$Z_c^2 = Z_{in} Z_l \tag{8}$$

$$Z_{in} = Z_c \frac{Z_l + jZ_c tg\beta l}{Z_c + jZ_l tg\beta l} \tag{9}$$

Assuming that the switch is in an ideal state, the insertion loss and isolation can be obtained by formula (10) [10], where y1 and y2 represent the normalized admittance of two ports and θ is the phase difference. In fact, the single-pole double-throw switch can not achieve the ideal turn-on and turn-off, so when calculating the attenuation of one port, the influence of the other port cannot be ignored. The two switches are each 1/4λ to the branch, so θ can be taken as 90°. Therefore, according to the formula (10), the insertion loss of the single port Port1 can be directly written as formula (11). Finally, the optimum position of the switch can be obtained by inserting the loss versus the admittance (the reciprocal of the impedance) and then the extremum of the distance l. The same algorithm can obtain the isolation. Figure 5 shows the structural model of the SPDT switch and the parameters in Figure5 are given in Table 2.

$$L_{dB} = 10\lg\frac{1}{4}|(2 + y_1+y_1y_2)\cos\theta + j(2 + y_1+y_1y_2)\sin\theta|^2 \tag{10}$$

$$L_{dB} = 10\lg\frac{1}{4}\left|\frac{1+2y_1+y_1y_2+y2}{2y1}\right|^2 \tag{11}$$

Table 2. Optimal configuration of SPDT.

Symbol	Describe	Value
W_1	width of 70.7Ω impedance transmission line	60μm
L_1	length of 70.7Ω impedance transmission line	130μm
L_2	length of 50Ω impedance transmission line	300μm
W	width of center transmission line	100μm
G	channel width	60μm

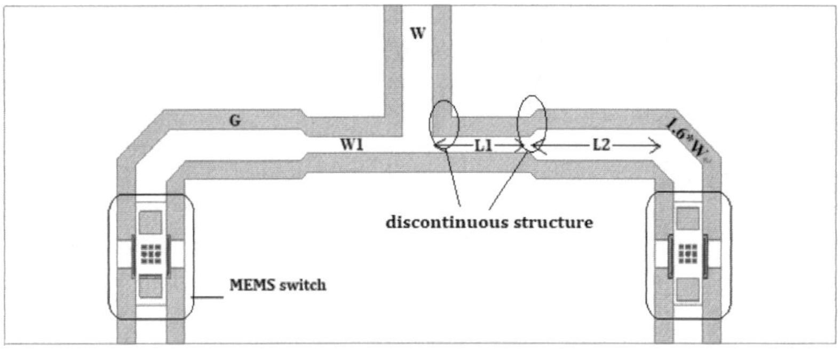

Figure 5. SPDT Switch Model.

5. Results Fabrication, Measurement, and Results

We then measure the model and observe the S-parameter results (as shown in Figure 7, 9). For SPST switches, the insertion loss is greater than -0.6dB when the switch is in the upper state, and the isolation is less than - 38dB when the switch is in the lower state at 35 GHz. For SPST switches, it can be seen that when one switch is in the upper state and the other switch is in the down state, the return loss is - 23dB in 35GHz. The insertion loss is -1dB and the isolation is less than -35dB.

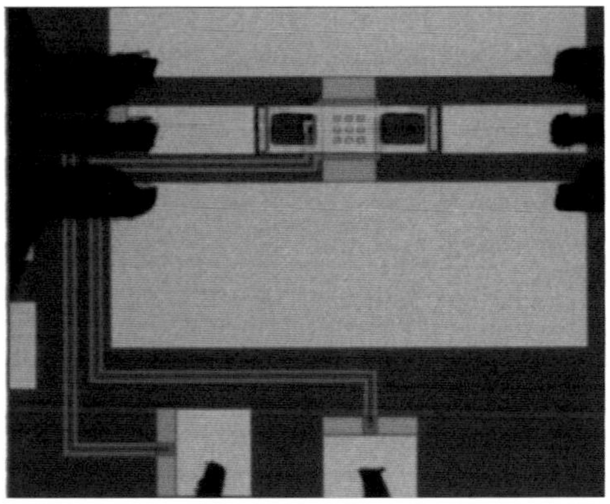

Figure 6. Processed MEMS switch.

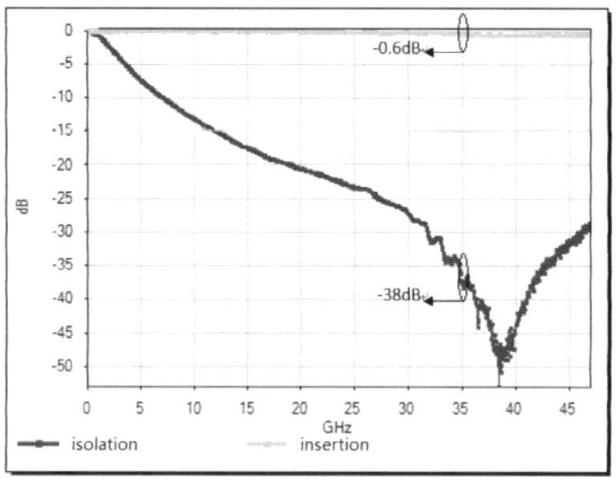

Figure 7. S parameters of MEMS switch.

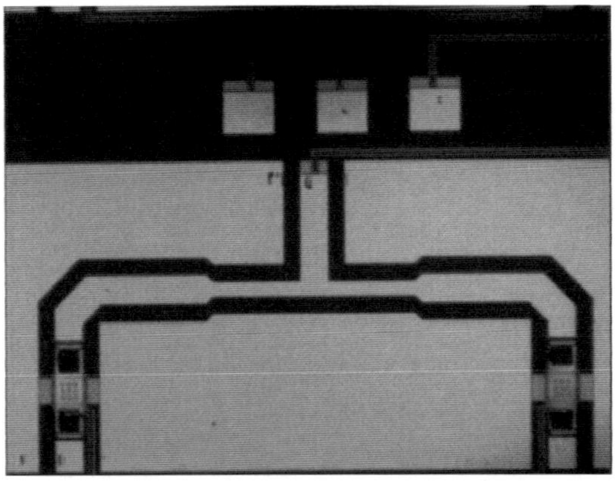

Figure 8. Processed SPDT switch.

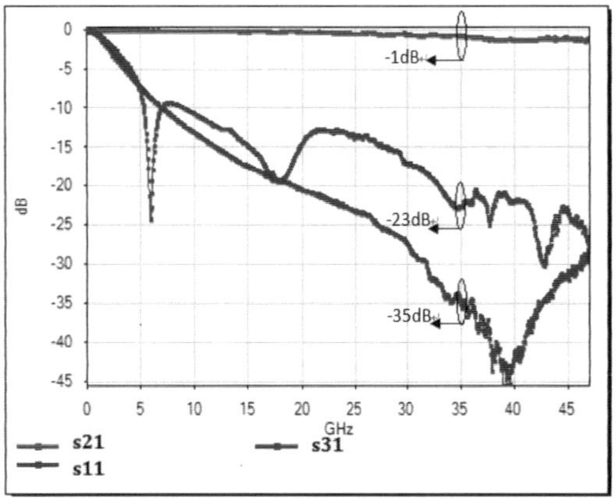

Figure 9. S parameters of SPDT switch.

6. Conclusion

This paper introduces the design and analysis of RF MEMS SPDT switches, and proposes a model structure sandwiched by two dielectric layers. This special structure can increase the contact area between the electrode and the beam, and can also make the capacitance ratio of the switch up and down state higher, which can make the switch get better performance. The SPDT switch selection function is realized by controlling the two switches of the T-type CPW to be in different states. Due to the small size and high performance characteristics, the proposed switch design can be used in the mode of reconstructing antenna feed network.

References

[1] G M Rebeiz, RF MEMS Theory, Design, and Technology *New York: Wiley* (2002)
[2] Caekenberghe K V RF MEMS on the radar *IEEE Microwave Magazine* **10**(6) 99-116 (2009)
[3] Daneshmand M and Mansour R R "RF MEMS Satellite Switch Matrices." *Microwave Magazine IEEE* **12**(5) 92-109 (2011)
[4] Liu Z W *et al.* "Design and measurement of RF-MEMS CPW on HRS." *Optics & Precision Engineering* (2005)
[5] Fouladi S andMansour R R "Capacitive RF MEMS Switches Fabricated in Standard 0.35-µm CMOS Technology" *IEEE Transaction on Microwave Theory and Techniques* **58**(2) 478-86 (2010)
[6] Dey Sukomal and Koul S K "Design and development of miniaturized high isolation MEMS SPDT switch for Ku-band T/R module application" *IEEE MTT-S International Microwave and RF Conference* (2013)
[7] Deepak Bansal *et al.* "Low Actuating Voltage Spring-Free RF MEMS SPDT Switch" *Journal of Electrical and Computer Engineering* 1-7 (2016)
[8] Y Gou *et al.* "Design and Simulation of a Novel RF MEMS SP2T Switch" *Journal of Microwaves s1* 345-8 (2010)
[9] David M Pozar *Microwave engineering* (2005)
[10] Jie Yan and L Xiao-Ping "Research of a X-Band MEMS Single-Pole-Double-Throw (SPDT) Switch" *Chinese Journal of Electron Devices* **29** 92-4 (2006)

IOP Conf. Series: Journal of Physics: Conf. Series **1209** (2019) 012013 doi:10.1088/1742-6596/1209/1/012013

An investigation into the high strength bonding technology of wafer-to-wafer with large-scale au line

S X Wang[1, *], J Zhu[1,2], S X Jia[1], Y Wu[1]

[1] Nanjing Electronic Devices Institute, No.524, zhongshan road,Nanjing, Jiangsu Province, China

[2]Science and Technology on Monolithic Integrate Circuits and Modules Laboratory, Nanjing Electronic Devices Institute, Nanjing City, Jiangsu Province, China

[*]Correspondence: wangshouxu@qq.mail.com; Tel.: +86-25-68005912

Abstract. As well known, the sandwich-like construction is the classic unit construction of MEMS microthrusters, and these three layers should be bonded between the igniting circuit layer and the microchamber-array-layer, the microchamber-array-layer and the micronozzle-array-layer. It's bonded by BCB between the igniting circuit layer and the microchamber-array-layer, and their bonding strength is the key factor to determine the reliability of MEMS microthruster-array-chips. And the failure is mostly discovered on the bonding surface of the igniting circuit layer and the microchamber-array-layer in ignition experiments. For solving the bonding failure, the failure reason will be analysed and the improved bonding technique is designed. Here, the bonding strength is externalized, and the average shear strength increase to 11.63MPa from 2.80Mpa, this is over 4times than before. The igniting experiments prove the Si3N4 layer can enhance the BCB bonding strength and this new bonding method does solve the bonding problem of MEMS microthruster-array-chips.

1. Introduction

Benzocyclobutene (BCB) has been gained comprehensive attention because of its unique properties such as low dielectric constant, low moisture absorption, low cure temperature, high degree of planarization, low level of ionic contaminants, high optical clarity, good thermal stability, excellent chemical resistance, and good compatibility with various metallization systems. Therefore, BCB is applied widely in bonding and packaging processes of 3D integrated circuits (IC) devices [1], optical devices [2], and micro electro mechanical system (MEMS) devices [3-8].

Bonding technology is one of the most important techniques in fabrications and packages of 3D IC devices and MEMS devices, and BCB is the key bonding materials in applications of wafer-to-wafer bonding, such as Silicon to Silicon and Silicon to glass. So, the BCB has been used as the bonding material between the micro igniters-array-chip and the micro chambers-array-chip in MEMS microthrusters-array-chips, which is the classic construction of like-sandwiches. As well known, the BCB could produce the great bonding strength with Si and compounds of silicon, but the bonding strength of BCB is not good enough with metal films. The bonding surface of igniters-array-chip is shown in Fig.1. However, there are a lot of Au lines on the bonding surface of micro igniters-array-chip. Therefore, the delamination [9] will be discovered on the bonding surface of igniters-array chips

IOP Conf. Series: Journal of Physics: Conf. Series **1209** (2019) 012013 doi:10.1088/1742-6596/1209/1/012013

and microchambers-array-chips. Equally, the fail phenomena of bonding surfaces have been observed in the igniting experiments of MEMS microthruster-array-chips. The representative bonding failure could be seen in Fig.2.

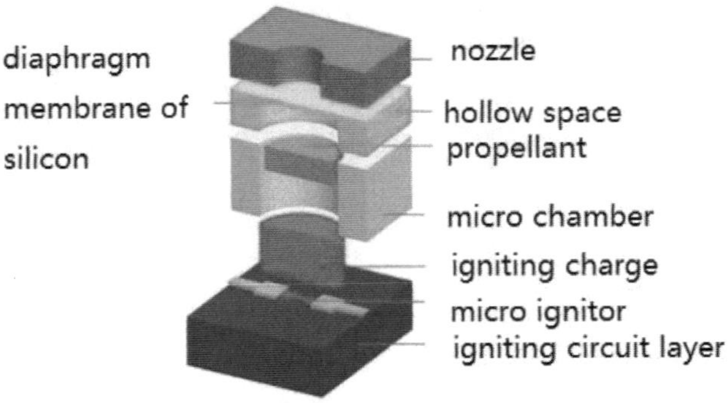

Figure 1. The unit construction of MEMS microthrusters.

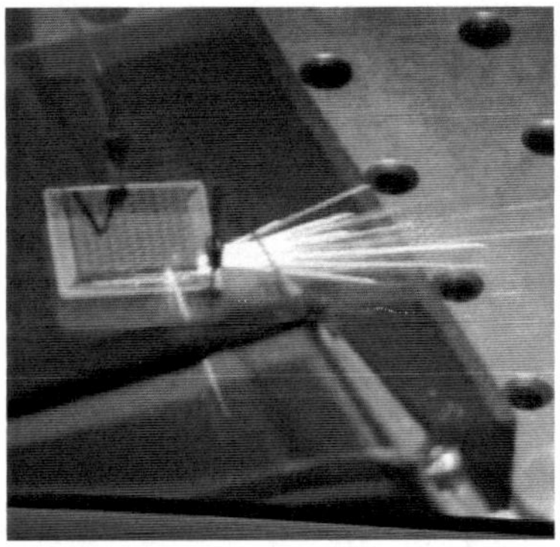

Figure 2. The classical failure in igniting experiment of MEMS microthruster-array-chips.

Above all, the bonding strength is one of the most important parameters which determines the working reliability of MMES microthrusters-array-chips.

In order to improve the bonding strength in fabrications and the reliability in applications of MEMS microthrusters, the dielectric layer has been applied in bonding process of MEMS microthruster-array-chips, and the bonding strength is enhanced, obviously.

2. Analysis into the reasons on bonding failure

2.1. The bonding mismatch on interfaces of BCB and igniting circuit layers

As well known, the module construction of BCB monomers is shown in fig.3 before its pre-hardening. The module construction of pre-hardening BCB shows that the BCB has great adhesion with materials of silicon, silica and silicon nitride, but BCB has not good adhesion with metallic materials, such as Au (the bonding surface of igniting circuit layer is shown in Fig.4). This is determined by the chemical properties of BCB and bonding materials, so some new method have to be developed.

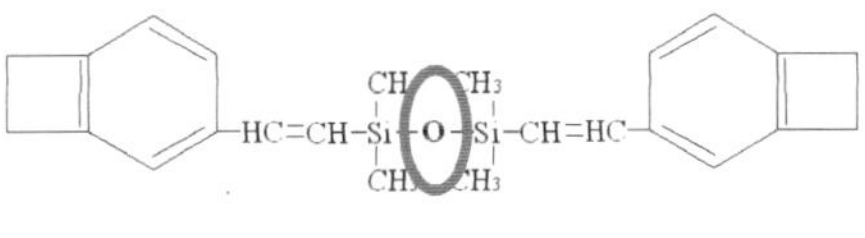

Figure 3. The module construction of BCB monomors.

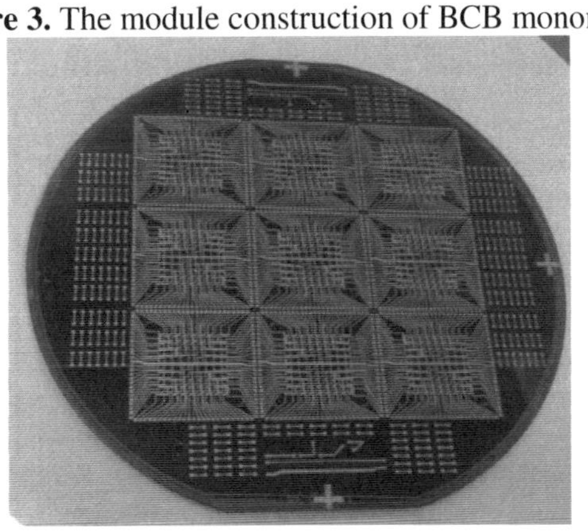

Figure 4. The igniting circuit layers.

IOP Conf. Series: Journal of Physics: Conf. Series **1209** (2019) 012013 doi:10.1088/1742-6596/1209/1/012013

2.2. Bonding failures

As the classic construction, the unit of MEMS microthruster has three layers: the igniters-array-layer, the microchamber-array-layer, and the micronozzle-array-layer. The fabricating process references Fig.5. And the bonding surface is coated by BCB film applied bladecoating processes with patterned film, before bonding the igniting circuit layer and the microchambers layer (shown in Fig.6).

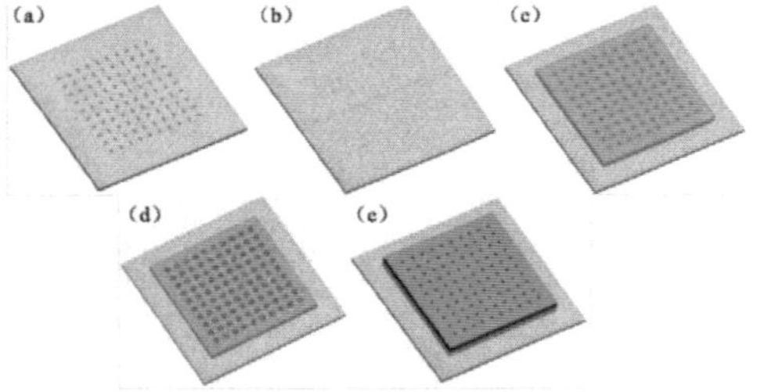

Figure 5. The fabricating process of MEMS mocrothruster-array-chips.

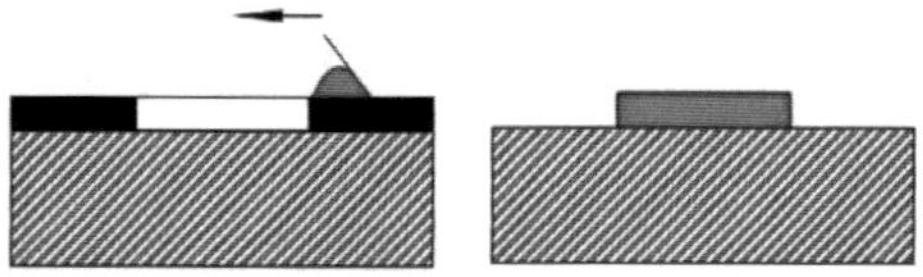

Figure 6. The bladecoating process of BCB in fabrications of MEMS microthruster-array-chips.

In another hand, there are large-scale Au lines on the bonding surface of igniting circuit layer. The thickness of Au lines is 3.2-3.5μm, and their minimum width is only 20μm. Therefore, it's difficult to blade coat the BCB into the spaces between Au lines, because of the high viscosity of BCB and the small width between Au lines. The bonding surface of igniting circuit layers are shown in Fig.4.

From the Fig.2, we could get the serious consequence in igniting experiment of MEMS microthruster-array-chips, and this failure of bonding process is an important problem what has to be solved.

In order to analysis the failure reasons in bonding process of MEMS microthruster-array-chips, the bonding surfaces are investigated about the BCB bonding layer and the surfaces of igniting circuit chips. we apply the shovel blade to separate the microchamber layer and the igniting circuit layer for investigate the bonding surfaces. The fig.7 shows how to separate the bonding layers. The bonding surface of igniting circuit layers is shown in Fig.8, and the bonding surfaces of microchamber-array-layer are shown in Fig.9. There are some micro hollows are discovered on the bonding interfaces of BCB, and these micro hollows obviously decreased the bonding strength, which is one of the key factors to determine the reliability of MEMS microthruster-array-chips in their tests and their applications in future. These micro hollows are generated when the BCB is coated on the bonding surfaces with coating and scrapping method, and these BCB films are not uniform enough. Therefore, the coating process of BCB should be improved for better bonding strength.

In fig.8 and fig.9, the defect zones are highlighted by red circles. They indicate the reasons why reduces the low bonding strength. And it's very hard to edulcorate these defect zones. Therefore, the new method is developed for decreasing these defect zones on the BCB layer and further for increasing the bonding strength between the igniting circuit layer and the microchamber-array-layer.

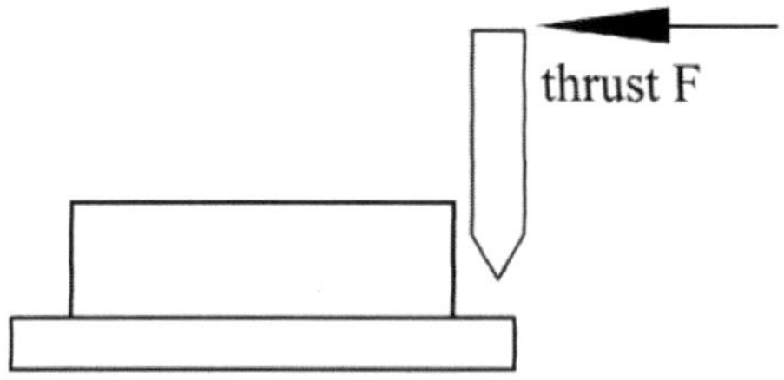

Figure 7. The illustration to separate the bonding layers.

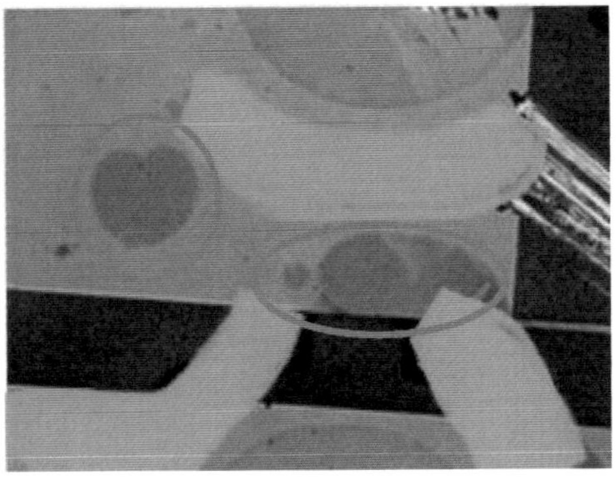

Figure 8. The bonding surface of igniting circuit layer.

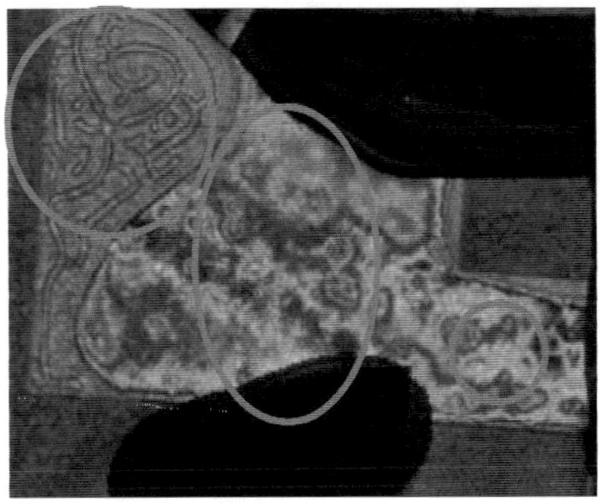

Figure 9. The bonding surface of microthruster-array-layer.

3. Results and Discussion

3.1. Adding thedielectrometric layer of silicon nitride on the surface of the igniting circuit layer

From the module construction of BCB monomer, the bonding strength is better when the BCB with silicon and inorganic silica materials (such as silicon, silicon nitride, and silicon oxide) than metallic materials (i.e. Au).

Therefore, the silicon nitride layer is deposited on the surface of Au line to enhance the adhesion after bonding process of igniting circuit layers and microchamber-array-layers.

3.2. To improve the patterning technique of BCB

The original patterning process of BCB is scraped on the bonding surface of igniting circuit layer with patterned film in fig.6. However, its bonding strength is not good enough, and there are some micro defects on the bonding interfaces of microchamber layers and igniting circuit layers, such as folds and micro holds of BCB membranes (in fig.8 and fig.9). Additionally, the thickness of BCB layers is not consistent with the blade-coating process.

In order to improve the thickness consistency of BCB bonding membrane, the blade-coating process is replaced by the spin-coating process. The pre-hardening BCB bonding membrane is patterned by the ICP (Inductively Coupled Plasma) etching process, and the patterned resistance is applied as the masking layer. The improved patterning process of BCB is shown in fig.10.

After the process of fig. 10, the microchamber-array-chips are bonded on the surfaces of igniting ciruit layers, what is shown in fig.11.

IOP Conf. Series: Journal of Physics: Conf. Series **1209** (2019) 012013 doi:10.1088/1742-6596/1209/1/012013

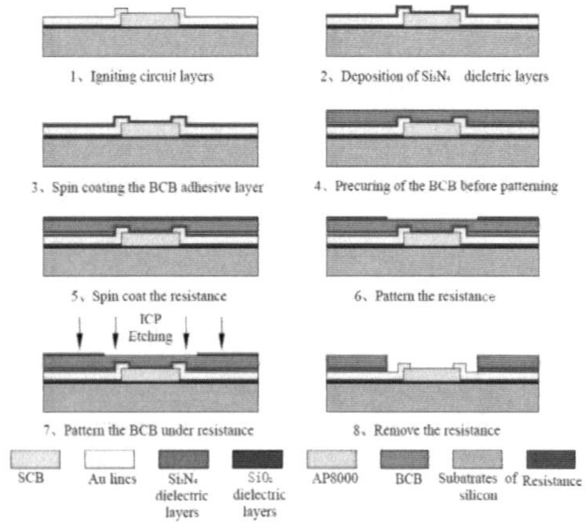

Figure 10. The improve patterning process of BCB.

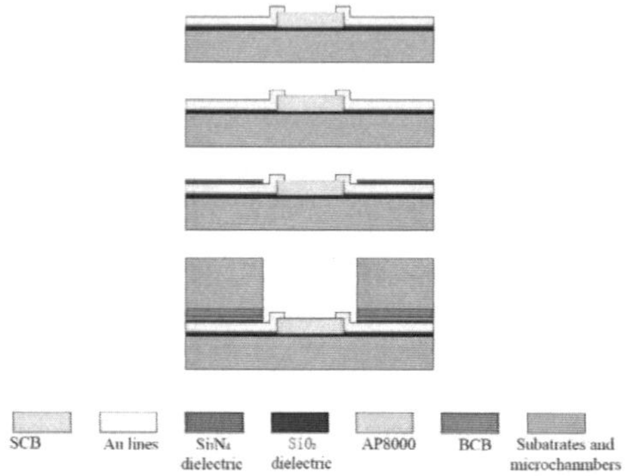

Figure 11. The advanced bonding process of igniting circuit layers and microchamber-array-layers.

3.3. Tests of shear strength after bonding of igniting circuit layers and microchamber-array-layers

After bonding the igniting circuit layer and microchamber-array-layers, a MEMS microthruster-array-chip are cut into nine pieces at the same area, and the shear strength is tested to characterize the bonding strength for contrast the advances of two bonding processes. The test results certify that the shear strength is over 4 times than before. The experimental results are shown in fig.12.

Finally, the igniting experiment demonstrates the new bonding technique increases the bonding strength, and this improved bonding method can ensure the reliability of MEMS microthruster-array-chips. The igniting test is shown in fig.13 on the combustion of MEMS microthrusters unit what is recorded by the high-speed optical camera, and the combustion experiment is shown in fig.14 on the MEMS microthruster-array-chips ignited by the control circuit unit.

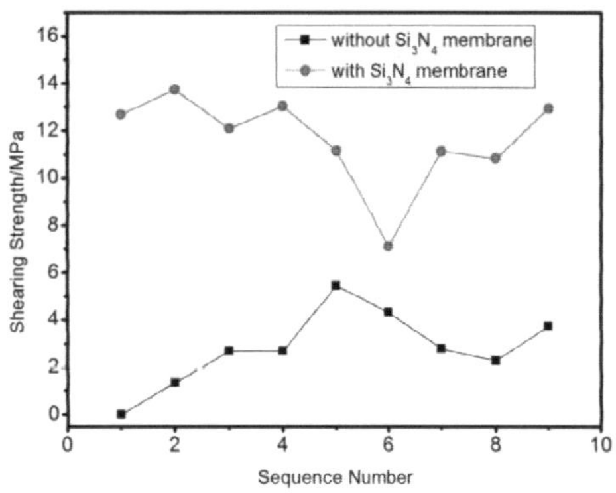

Figure 12. Contrasts of bonding strength between two bonding techniques.

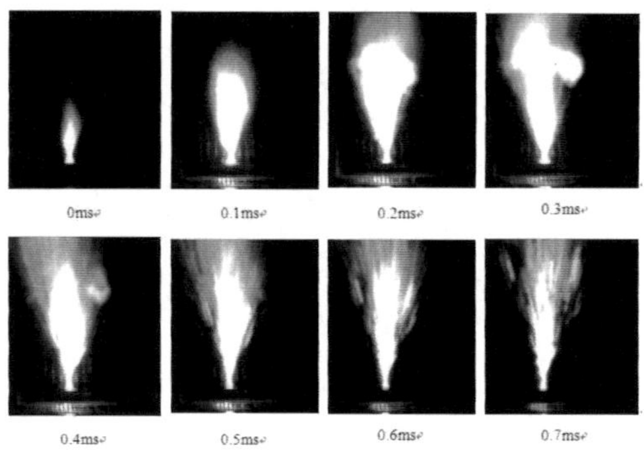

Figure 13. The combustion process of a MEMS microthruster unit recorded by high-speed optical cameras.

Figure 14. Igniting tests of MEMS microthruster-array-chips ignited by the igniting control unit.

4. Conclusion

For increasing the bonding strength, a new bonding technology is created by adding the patterned silicon nitride layer on the surface of metallic lines, and the knife coating technique is replaced by the patterning technology on the BCB adhesive layers.

Contrast the shearing tests, the experimental results indicated that the new bonding process could increase the bonding strength, obviously. The bonding strength of new bonding technique is over four times than the original knife coating process.

Above all, this investigation reveals a new bonding technology, and it can be applied in wafer-to-wafer bonding, single-chip bonding, multi-bonding and 3D bonding and packaging, comprehensively.

Acknowledgments

No.

References

[1]S.L. Lin, W.C. Huang, C.T. Ko, K.N. Chen. BCB-to-oxide bonding technology for 3D integration, *Microelectronic Reliability*, 52 (2012) pp.352-355.

[2]Andy Zhenzhong Zhang, Qin Wang, Stefan Karlsson,etc. Fabrication of an electro-absorption transceiver with a monolithically integrated optical amplifier for fiber transmission of 40-60 Ghz radio signals[J], *Semiconductor Science and Technology*,16 (2011) 014042(6pp)

[3]Lei Liu, Minghao Zhang, Susu Li, Bo Chen. Wafer-level Adhesive Bonding with Benzocyclobutene for the Acceletrometer[J],*Eletronic Science and Technology*, 9 (2012) 9-12.

[4]Zhihui Liu, Lei Tian, Yuling Li, Yanzhao Yin. Research on BCB bonding process of acceleration sensitive chip made of three-layer Si wafer[J], *Transducer and Microsystem Technologies*, 3 (2015) 37-43

[5]Yufei Liu, Wenping Liu, Sihua Li, Yaming Wu, Le Luo. Si-Glass Hermetic Package with the Benzo-Cyclo-Butene Material[J], *Chinese Journal of Semiconductors*, 27 (2006) 407-410.

[6]N. Lorenz, M.D. Smith D.P. Hand. Wafer-level packaging of silicon to glass with a BCB intermediate layer using localised laser heating[J], *Microelectronics Reliability*, 51 (2012) 2257-2262.

[7] Lubing zhao, Jing Xu, Shaolong Zhong, Shaoliang Li, Yaming Wu. Research on Si-si bonding technique with photosensitive BCB for MEMS devices[J], *Transducer and Microsystem Technologies*, 8 (2014) 48-51.

[8] K.O.Lee. Thermomechanical Reliability study of Benzocyclobutene Film in Wafer-Level Chip-Size Package[J], *Journal of Electronic Materials*， 4 (2012) 706-711.

[9] Fan Bu, QingMa, Zheyao Wang. Delamination of bonding interface between benzocyclobutene (BCB) and silicon dioxide/silicon nitride[J], *Microelectronics Reliability*,2016:65, pp.225-233.

IOP Conf. Series: Journal of Physics: Conf. Series **1209** (2019) 012014 doi:10.1088/1742-6596/1209/1/012014

Simulation of explosion proof capacity of material partition in micro initiation sequence

LI Hui[1], REN Wei[1,2], BAI Ying Wei[1] and LIU Tian[1]

[1] Science and Technology on Applied Physical Chemistry Laboratory, Shaanxi Applied Physics-Chemistry Research Institute, Xi'an, 710061, China
[2] The State Key Laboratory for Manufacturing Systems Engineering, Xi'an Jiaotong University, Xi'an, 710049, China
Abstract. In order to study the effect of materials gap on detonation capability of micro sequence. The law of the micro sequence initiating ability is simulated using the finite element analysis method in the different material partition. The results show that the capacity of detonation wave pressure attenuated by Ni partition is stronger than the steel. Then the simulation results are tested and verified, indicating that the simulation of initiation sequence can reflect the actual situation accurately.

1. Intorduction

MEMS initiating explosive device is becoming the key technology for the development of the new generation of intelligent ammunition[1]. Miniaturized detonation transfer sequence based on MEMS initiating technology is an important technology to realize the safety of MEMS and initiating devices. The baffle interface in the detonation transfer sequence is a typical energy transfer interface. By changing the thickness of the baffle, the detonation transmission or partition between the upper and lower charge can be realized, the reliability and safety of the explosion sequence are improved at the same time. After the small diameter charge is detonated, the detonation wave is absorbed at the interface of the partition, then it spreads inside the partition, and it decreases as the propagation distance increases. In addition, the attenuation coefficient of the shock wave in the separator varies with the material of the separator[2]. At present, the donor explosive used by Hu Xiangyu[3] is 8701 explosive with a diameter of 50 mm. Wang Zuoshan[4] used RDX with a diameter of 5.1 mm, F.X. Jette[5] used nitromethane with a diameter of 50 mm, and they obtained the attenuation law of shock wave in the plexiglass separator respectively. But it is mainly the detonation output of large diameter charge.

The geometry of the micro-initiation sequence studied in this paper is shown in figure 1. CL-20 and JO-9C are used as the first and second stages of the detonation sequence, and the dimensions are 2*2 mm and 4*4 mm respectively. Based on Lee-Tarver ignition growth model, the influence of stainless steel and nickel separator on the detonation capability of micro-sequences are analyzed by finite element simulation method, and are verified by experiments. The research results have important reference value for further research on the structural design of the micro-initiation series.

IOP Conf. Series: Journal of Physics: Conf. Series **1209** (2019) 012014 doi:10.1088/1742-6596/1209/1/012014

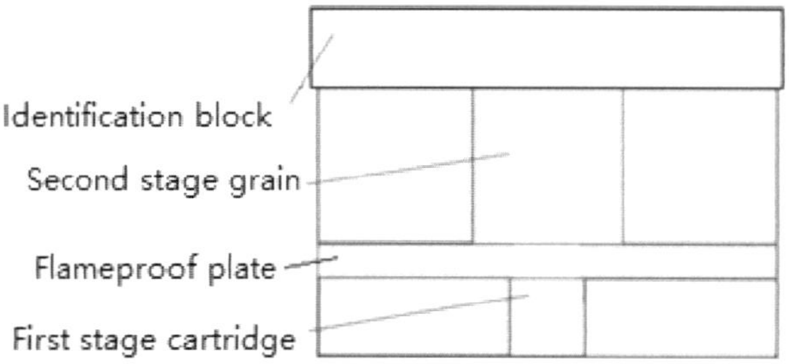

Figure 1. Structure of micro sequence.

2. Numerical simulation of initiation of micro sequence

2.1. Establishment of micro sequence initiation model

2.1.1. Finite element model. The initiation of micro sequence is simulated by the finite element method. AutoDyn software is selected for calculation[4]. The model is composed by four parts: a first-stage explosive column, the separator, a second-stage explosive column, and an identification block. The materials are CL-20, steel/nickel sequence, JO-9C and aluminum identification blocks respectively. The micro sequence structure is a rotating body, so its 1/2 model is established due to its axis symmetry, as shown in figure 2.

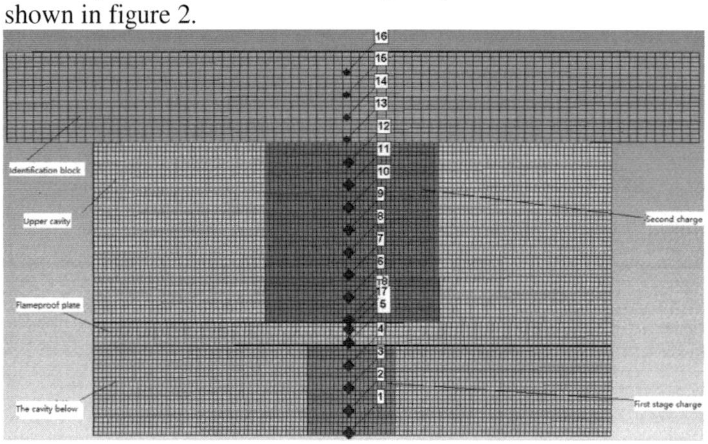

Figure 2. The simulation model.

In order to simulate effectively, the model is simplified reasonably:

(1) the initial stress of all materials in the micro initiation sequence is 0. The explosive is restrained by the shell, and the rest of the boundary is free surface.

(2)The density of the explosive column is uniform. Micro detonators were not considered in the simulation. The simulation begins with the first stage of the explosive column, and its center is detonated. In order to obtain the pressure-time curve, observation points are established at the symmetry axis position of the model. It is judged whether the two columns are detonated by the pressure curve of the observation point.

2.1.2. State equation and material parameters. The JWL equation describes the explosion process of the first stage explosive cartridge in the following form [6-7]:

$$P = A(1-\frac{\omega}{R_1 V})^{e^{-R_1 V}} + B_e(1-\frac{\omega}{R_2 V})^{e^{-R_2 V}} + \frac{\omega E}{V} \quad (1)$$

Among them, P is pressure. V is relative volume. E is energy density. A, B, R_1 and R_2 are constants. The value of constants are determined by fitting the data of Hugoniot and initial sound velocity.

Table 1. The simulation parameters of JWL equation.

ρ_0 (g/cm³)	A(Mbar)	B(Mbar)	R_1	R_2	ω	E/V（Gerg/mm）
1.92	16.37	1.86	6.5	2.7	0.5	0.115

The Grunessen equation of state is used for the partition material and the identification block. It is expressed by the compression pressure of the compressed material and the expansion pressure of the expanded material. The compression pressure is defined as:

$$P = \frac{\rho_0 C^2 \mu \left[1 + (1 - \frac{\gamma_0}{2})\mu - \frac{a}{2}\right]}{\left[1 - (S_1 - 1)\mu - S_2 \frac{\mu^2}{\mu + 1} - S_3 \frac{\mu^3}{(\mu + 1)^2}\right]} + (\gamma_0 + a\mu)E \tag{2A}$$

Among them, $\mu = 1/V - 1$, $u_s = C_0 + Su_p$

The expansion pressure is defined as:

$$p = \rho_0 C^2 \mu + (\gamma_0 + a\mu)E \tag{2B}$$

In the formula, C is the intercept between the velocity of the shock wave and the particle velocity curve. Gamma₀ is the Gruneisen coefficient. A is the first order volume correction of gamma₀. S_1, S_2 and S_3 are the slopes of the shock velocity and particle velocity curves. It is the degree of compression. V is the relative volume; u_s is the speed of shock wave; u_p is the velocity of protons (particles); C_0 is the speed of sound.

Table 2. The simulation parameters of Gruneisen equation.

Material type	ρ_0 (g/cm³)	C(s/km)	S_1	S_2	S_3	ω	a
steel	7.83	4.569	1.490	0	0	2.17	0.46
nklc	8.81	4.19	1.54	0	0	2.14	0.41

The second stage explosive column is an explosion caused by shock wave pressure. It is a non-ideal detonation, so its reaction process cannot be ignored. The Lee-Tarver equation is used to describe its action process[8-10].

$$\frac{dF}{dt} = I(1 - F)^b(\frac{\rho}{\rho_0} - 1 - a)^x + G_1(1 - F)^c F^d p^y + G_2(1 - F)^e F^g P^z \tag{3}$$

F is the degree of reaction of the explosive. T is time and ρ is density. ρ_0 is the initial density. P is the pressure. The equation consists of 12 unknown parameters: I, B, a, x, G1, C, D, y, G_2, e, G and Z.

Table 3. The simulation parameters of Lee-Tarver equation.

ρ_0 (g/cm³)	A(Mbar)	B(Mbar)	R1	R2	ω	E/V（Gerg/mm）
1.84	8.524	0.108	4.6	1.3	0.38	0.102
I(μs-1)	a	b	G1(Mbar-zμs^{-1})	c	d	y
7.43e+1011	0.667	0	3.1	0.667	0.111	1
G2(Mbar-zμs-1)	e	g	z	Figure max	F_{G1max}	F_{G2max}
400	0.333	1.0	2.0	0.3	0.5	0.0

3. Simulation and analysis of micro-sequence flameproof

The fluid-solid coupling method is used to simulate the explosion process of the micro-explosion sequence. When the first stage explosive column is completely exploded, the pressure on the input

IOP Conf. Series: Journal of Physics: Conf. Series **1209** (2019) 012014 doi:10.1088/1742-6596/1209/1/012014

surface of the second-stage explosive column is as shown in figure 3, after the shock wave passes through the separator of stainless steel and nickel material of different thicknesses.

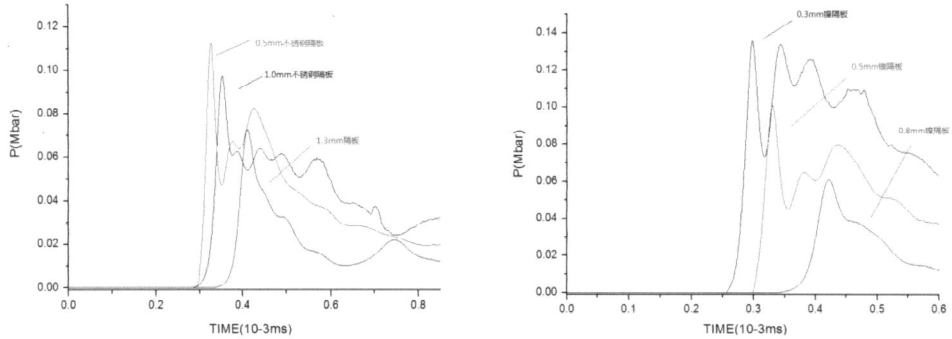

(a) Steel layer (b) Ni layer

Figure 3. P-T curve of the input surface of the second explosive column through steel and nickel separators of different thicknesses.

As shown in figure 3, When both the stainless steel separator and the nickel separator have a thickness of 0.5 mm, the detonation wave passes through the steel baffle to the input of the second stage explosive column at 0.3 s, and the maximum pressure is 11.85Gpa. The detonation wave passes through the nickel separator to the input of the second stage explosive column at 0.35 s, and the maximum pressure is 9.8Gpa. It can be seen that the detonation wave propagates faster in the stainless steel separator than in the nickel separator. And nickel is more resistant to detonation pressure than stainless steel.

3.1. Simulation results of detonation after stainless steel partitions

The detonation process of the second-stage explosive column after the different thickness of the stainless steel separator is simulated. The effect of stainless steel on the propagation of detonation waves was studied. The observation point represents the height in the explosive column. The higher the value of the observation point, the higher the explosive column and the farther from the input end. The pressure-time curves at different observation points represent changes in pressure at different heights of the cylinder over time. The pressure-time curve indicates the detonation results of the second stage explosive column. The result is shown in figure 4.

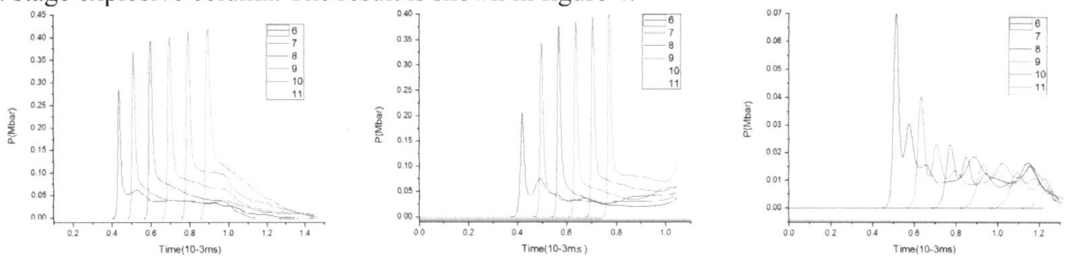

(a) 0.5mm thickness (b) 1.0mm thickness (c) 1.3mm thickness

Figure 4. P-T curve of the second-stage explosive column when the partition material is steel.

As shown in figure 4(a), when the thickness of the separator is 0.5 mm, the detonation wave propagates to the input of the second stage explosive column at 0.3 s, and the second stage explosive column is detonated. As time goes by, the pressure increases quickly. The pressure is greatest and the pressure reaches a steady value at 0.53 s. The results show that the second stage explosive column can achieve complete detonation. In figure 4(b), when the thickness of the partition is 1.0 mm, when the detonation wave propagates to the input end of the second-stage column, the peak value of the pressure at the observation point gradually increases along the direction of the propagation of the detonation wave. Then the pressure is stable. The second stage explosive column achieves detonation. As shown in figure 4(b), when the separator thickness was 1.3 mm, the second-stage explosive column was not detonated. When the explosion propagates to the input of the second stage explosive column, the peak of the pressure at the observation point gradually increases to 7 Gpa in the direction of the detonation wave propagation. The pressure value then decreases with time. As the height of the explosive column increases, the peak value of the pressure at the observation point decreases significantly. Over time, the pressure peak is much lower than the detonation pressure.

3.2. Simulation results of detonation after nickel partitions

This part simulates the detonation process of the second stage explosive column after the detonation wave passes through the nickel separator of different thickness. The effect of nickel materials on the propagation of detonation waves is studied. The detonation of the second stage explosive column is obtained from the pressure-time curve. The result is shown in figure 5.

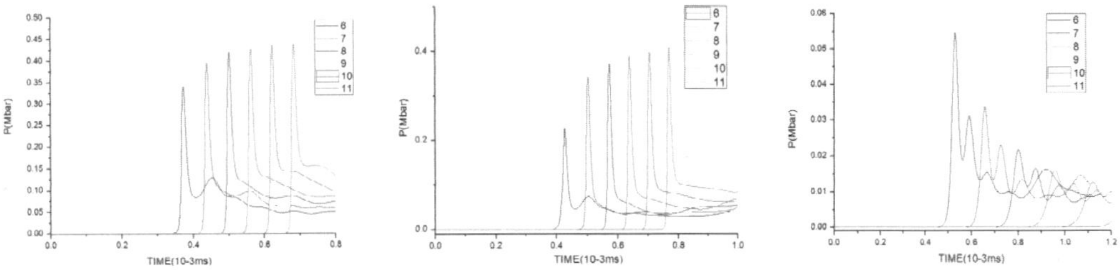

(a) 0.5mm thickness (b) 1.0mm thickness (c) 1.3mm thickness

Figure 5. P-T curve of the second-stage explosive column when the partition material is Ni.

As shown in figure 5, when the thickness of the separator is 0.3 mm and 0.5 mm, the pressure peak at the observation point gradually increases, along the direction of the detonation wave, and then reaches stability. The second stage explosives column detonated. When the thickness of the separator is 0.8 mm, the detonation wave is applied to the input end of the second-stage explosive column at 0.43 s. Along the direction of the detonation wave, the peak of the explosive column pressure rises to 5.6 Gpa, and then the pressure value decreases with time. As the height of the explosive column increases, the peak pressure at the observation point is significantly reduced. And over time, the pressure peak is much lower than the detonation pressure. The second stage explosives column was not detonated.

4. Micro-sequence explosion test verification

4.1. Test structure

In order to verify the simulation results, the ignition test was performed on the micro-initiation sequence. The test assembly structure is consistent with figure 1. Test samples and assembly drawings are shown in figure 6.

Figure 6. Testing device.

4.2. Test result

(1) When the material is stainless steel, the experimental results are shown in figure 7.

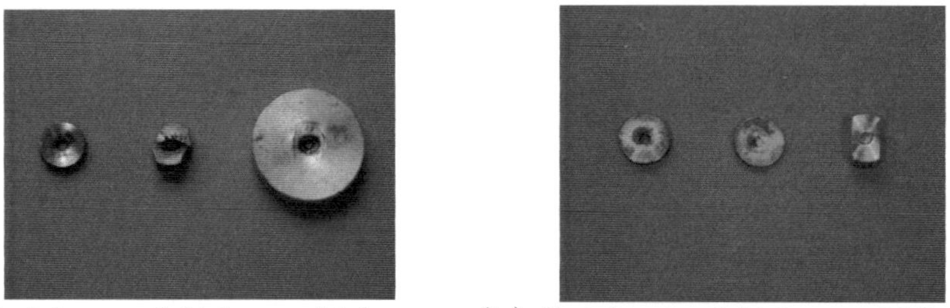

(a) Stainless steel thickness of 1.0mm (b) Stainless steel thickness of 1.3mm

Figure 7. Test results of detonation performance when the separator is stainless steel.

(2) When the material is nickel, the experimental results are shown in figure 8.

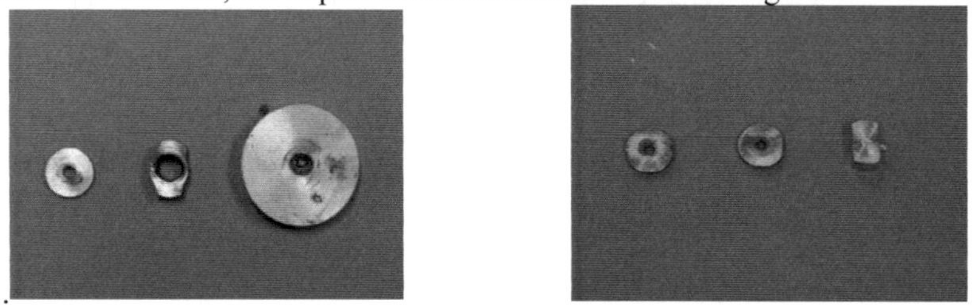

(a) Ni with a thickness of 0.5 mm (b) Ni with a thickness of 0.8 mm

Figure 8. Test results of detonation performance when the separator is Ni.

As shown in figure 7, when the thickness of the stainless steel separator is 1mm, the cavity containing the second stage explosives was cracked. there is a large indentation on the identification block. The results show that the JO-9C is completely detonated. When the thickness is 1.3mm, the partition is not bad, the second stage explosive column and cavity are intact. It indicates that the JO-9C has not exploded. In figure 8, when the thickness of the nickel separator is 0.5 mm, the cavity containing the second stage explosive column is broken, there is a large indentation on the identification block. It explains that the JO-9C is completely detonated. When the thickness is 0.8 mm, the partition is broken. However, there is no major damage to the second stage explosives column. There is not explosion in the JO-9C. The experitental results show that the simulation results are basically reliable.

5. Conclusions

Through the simulation research and experimental verification of the detonation wave transmission of the micro-initiation sequence of different separator materials, the following conclusions can be drawn:

(1) When stainless steel and nickel materials are used as the separator, the ability of the detonation wave to make the JO-9C explode is: stainless steel > nickel. Because the nickel separator has a strong attenuation effect on the detonation wave, for the above two materials, only from the perspective of achieving micro-sequence explosion-proof, nickel is a preferred material.

(2) The detonation ability of the explosion shock wave after being attenuated by the diaphragm is positively correlated with the pressure value at the input end of the second-stage explosive column. Although the effects of the initial pressure value and the pressure decay rate are only indirect, they are directly related to the residual pressure value. From the point of view of achieving micro-sequence explosion-proof, it is better to select materials with small initial pressure value and faster pressure decay under the same conditions.

References

[1]Kong Junfeng, Li Bing. A new pyrotechnic technology and the application. *Base Defense Technology*. 2010(7): 40-51

[2]Zhao Hai-xia, Xu Xin-chun. Attenuation Model of Shock Wave in Different Materials Gap. *Chinese Journal of Explosives & Propellesnys*. 2011, 34(6)

[3]WANG Zuo-shan, LIU Yu-cun,etal.Study on the attenuating model of detonation shock wave in

the PMMA gap[J]. *Journal of Basic Science and Engineering* ,2001,9(4):316-319

[4] Jette F X, Goroshin S, Higgins A J. Shock reactivity of non-porous mixtures of manganese sulfur[J]. *Shock Compression of Condensed matter*,2007:1033-1036

[5] WANG Hui, SHEN Fei, YUAN Jian fei, ZHANG Lei. Research on the Hugoniot Experiment and EOS of Product for JO-9159 Explosive Overdriven Detonation. *Chinese Journal of Explosives & Propellesnys*, 2014, 37(6): 12-15

[6] NAN Yu xiang, JIANG Jian wei, WANG Shu you, CHEN Dong ping,etal. JWL Equation of State of Detonation Product for CL-20 Based Pressed Composite Explosive[J]. *Chinese Journal of Energetic Materials*, 2015, 23(6):516-521

[7] SHEN Fei, WANG Hui, YUAN Jian fei. A Simple mentod for determining parameters of JWL EOS. *Journal of Vibration and Shock*, 2014:33(9)107-110

[8] SUN Zhan-Feng,LI Qing-Zhong,ZHANG Chong-Yu. Further Study on JWL Equation of State of Detonation Product for Insensitive High Explosive. *Shock Wave and Detonation Physics*. 2010:24(1)55-60

[9] E. L. Lee, C. M. Tarver. Phenomenological model of shock initiation in heterogeneous explosives. *Phys Fluids*. Vol23, No12, 1980: 2362-2372

Non-isothermal decomposition kinetics of copper azide nanowire array

Yanlan Wang, Fang Zhang, Lei Zhang, Ruishan Han, Rui Zhang and Jianhua Chen

Science and Technology on Applied Physical Chemistry Laboratory, Shaanxi Applied Physics-Chemistry Research Institute, Xi'an 710061, Shaanxi, China

E-mail:cloudyanlan@163.com

Abstract. Copper azide nanowire array (CA-NW) was a new primary explosive with high orientation. Thermal decomposition behaviour of CA-NW was studied to evaluate its thermal safety. Thermal decomposition reaction was observed by DSC and TG-DTG methods. Kissinger's and Ozawa's methods were used to obtain the kinetic parameters, such as apparent activation energy (E_k and E_o) and pre-exponential factor (A_k). The mechanism functions of thermal decomposition, self-accelerating decomposition temperature (T_{SADT}), critical temperature of thermal explosion (T_{bpo}) were obtained. Results show that for CA-NW, E_k= 94.08kJ·mol^{-1}, A_k=8.32×10^{7}min^{-1}, E_o=97.09 kJ·mol^{-1}, T_{SADT}=456.6K, and T_{bpo}=476.6K.

Keywords: copper azide, nanowire arrays, non-isothermal decomposition, thermal safety

1. Introduction

Copper azide is a kind of primary explosive with excellent performance, because it has the advantages of small input energy, large output energy, green environmental protection and so on [1-3]. However, its application is limited due to its extreme sensitivity to shock and friction [1].

There are two promising ways for copper azide to get practical application [2-8]. One way to do this is to tweak its sensitivity properties by tweaking its nanostructure, and another possible method is to isolate dangerous operations from operators by in-situ reaction.

Different nanostructure of copper, such as copper film, porous copper, copper nanowire and copper nano particulars confined inside CNTs, have been used to in-situ react with hydrazoic gas by our group [9-14].All of these four methods are MEMS-compatible processing techniques, which can realize the nano-crystallization and in-situ reaction of copper azide and fabrication of the energetic material on a chip.

Because of its highly oriented in one direction, copper azide nanowire arrays (CA-NW) is different in performance compared with other different structure of copper azide. Further research on the relationship between structure and performance is needed. In this paper, research efforts are ongoing to study thermal decomposition behavior of CA-NW. This study can help evaluate the application possibility of CA-NW.

2. Materials and Methods

IOP Conf. Series: Journal of Physics: Conf. Series **1209** (2019) 012015 doi:10.1088/1742-6596/1209/1/012015

2.1 Fabrication of CA-NW

CA-NW was prepared according to Ref. [9] and its morphology is shown in Figure 1. As clearly observed in Figure 1, CA-NW is highly ordered in length, approximately 500nm in diameter and 50μm in length, and the surface of copper azide is covered with convex structures. XRD result (Figure.2) confirms that CA-NW possess an orthorhombic structure. Except cuprite azide, there are also a little bit tetragonal cuprous azide and cubic copper diffraction peaks observed, which may be caused by the insufficient reaction time.

Figure 1. SEM image of CA-NW.

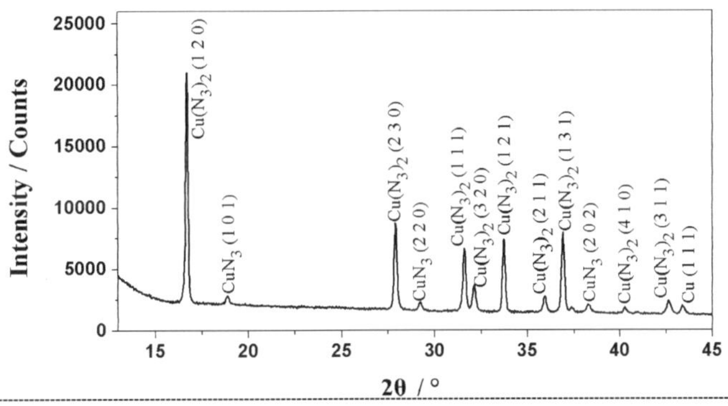

Figure 2. XRD pattern of CA-NW.

2.2 Apparatus and experiment parameters

A STA449F3 Simultaneous Thermal Analyzer (NETZSCH, Germany) was used for DSC experiment, the mass of sample was about 0.2mg. The experiment was carried out under the atmosphere of nitrogen, and the flow rate of nitrogen was 50mL·min^{-1}. The heating rate was 10.0K·min^{-1}.

The TG-DTG experiments were carried out by the same equipment, the mass of sample was about 0.2mg, The experiment was carried out under the atmosphere of nitrogen, and the flow rate of nitrogen was 50mL·min^{-1}.The heating rates were 2.5, 5.0, 10.0, 15.0K·min^{-1}, and the temperature conditions range from room temperature to 533K.

3. Results and Discussion

3.1 Thermal analysis

The DSC curve of CA-NW at a heating rate of 10K·min^{-1} is shown in Figure 3. The corresponding TG-DTG curve is shown in Figure 4. The results show that CA-NW has a strong exothermic peak in the range of 438K ~ 528K, and the mass loss is about 19.4%. The onset temperature T_o=452.8K, peak temperature T_p=488.8K, and ending temperature T_e =516.8K, respectively.

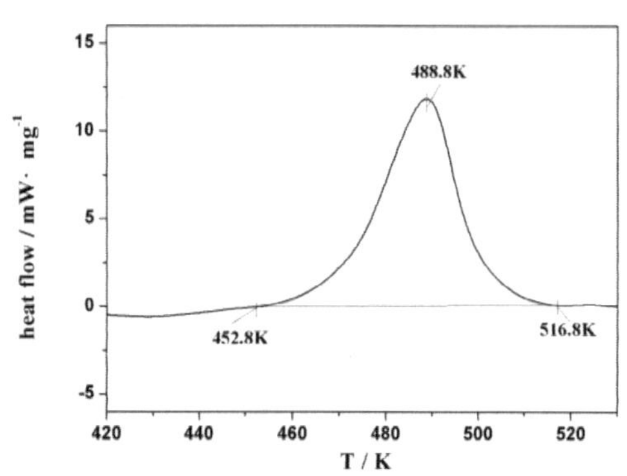

Figure 3. DSC curve of CA-NW.

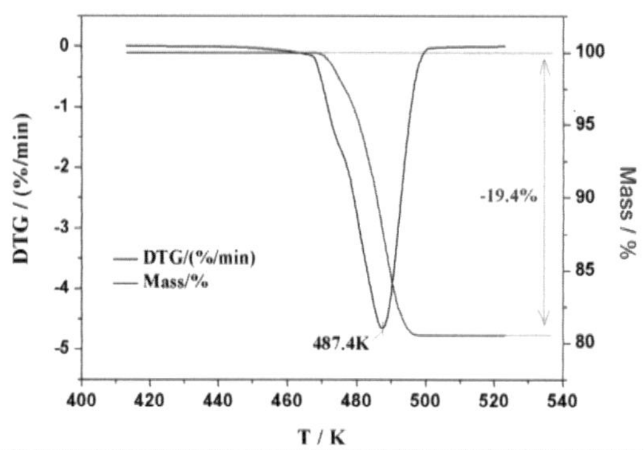

Figure 4. TG-DTG curve of CA-NW.

3.2 Non-isothermal decomposition kinetics of CA-NW

Figure 5 shows the CA-NW's DSC curves of curves in four different heating rates.

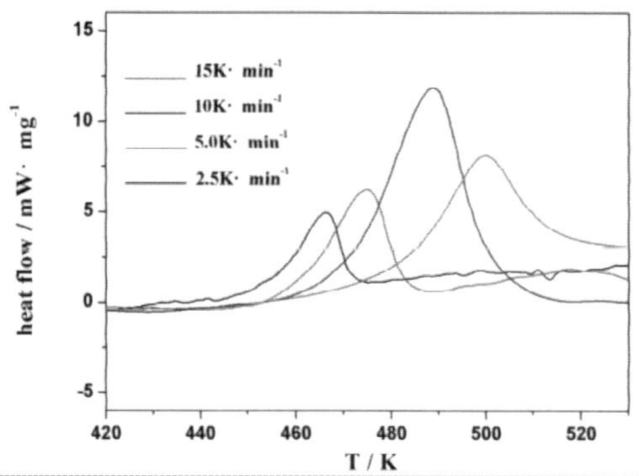

Figure 5. DSC curve of CA-NW at different heating rates.

The characteristic values, T_o, T_p and T_e, which were obtained under four different heating rates are listed in Table 1.

The kinetic parameters of exothermic decomposition of CA-NW were obtained by using Kissinger's method [15] and Ozawa's method [16], as shown in Equation 1 and Equation 2.

Kissinger's method:

$$\ln\left(\frac{\beta_i}{T_p^2}\right) = \ln\frac{A_k R}{E_k} - \frac{E_k}{RT_p},\ i = 1,2,3,4 \quad (1)$$

Where β_i is heating rate, $K{\cdot}min^{-1}$; T_p is peak temperature, K; E_k is apparent activation energy obtained by Kissinger's method, $kJ{\cdot}mol^{-1}$; A_k is pre-exponential constant, min^{-1}; R is ideal gas constant, 8.314 $J{\cdot}K^{-1}{\cdot}mol^{-1}$.

Ozawa's method:

$$\lg\beta_i = lg\left(\frac{AE_o}{RG(\alpha)}\right) - 2.315 - 0.4567\frac{E_o}{RT_p},\ i = 1,2,3,4 \quad (2)$$

Where E_o is apparent activation energy obtained by Ozawa's method, $kJ{\cdot}mol^{-1}$; G (α) is kinetic model function.

Table 1 lists the calculated values of kinetic parameters. E_k is consistent with E_o. The value of r_k and r_o (the linear correlation coefficients obtained by Kissinger's method and Ozawa's method) are all close to 1.

The calculation formula of T_{p0}(corresponding to $\beta\rightarrow0$) is shown in Equation 3, and the values is also listed in Table.1.

$$T_{pi} = T_{p0} + b\beta_i + c\beta_i^2 + d\beta_i^3,\ i = 1,2,3,4 \quad (3)$$

Table 1. Kinetic parameters from the exothermic decomposition reaction for CA-NW at various heating rates.

β $K{\cdot}min^{-1}$	T_o K	T_p K	T_e K	E_k $kJ{\cdot}mol^{-1}$	A_k min^{-1}	r_k	E_o $kJ{\cdot}mol^{-1}$	r_o	T_{p0} K
2.5	443.1	466.4	477.5	94.08	8.32E+7	0.99	97.09	0.99	456.6
5.0	448.5	474.9	489.6						
10.0	452.8	488.8	516.8						
15.0	455.0	499.9	529.6						

Fig.6 shows the temperature vs conversion rate (α) curve at different heating rates. Figure 7 shows the value of E calculated at a given conversion rate.

When the conversion rate is between 0.1 to 0.85, the value of E_o is stably distributed between 101.72 $kJ{\cdot}mol^{-1}$ to 80.91$kJ{\cdot}mol^{-1}$. The average value of E_o calculated is 92.04$kJ{\cdot}mol^{-1}$. The calculated value corresponds to the values obtained using the two methods mentioned above. So, this value is used in later studies to verify the reliability of other methods.

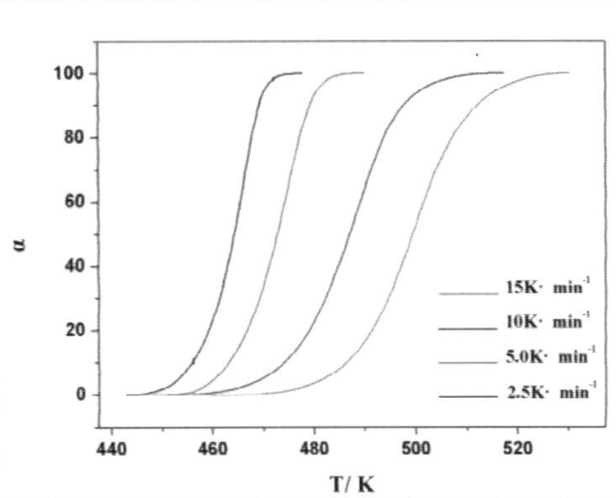

Figure 6. T vs α for the decomposition reaction of CA-NW at different heating rates.

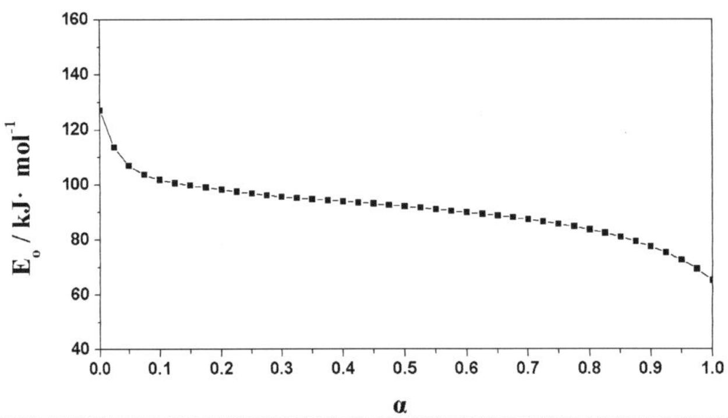

Figure 7. E vs α for the decomposition reaction of CA-NW at different heating rates by Ozawa's method.

The linear regression analysis of the corresponding E, A and G (α) (the most probable kinetic model function) from each DSC curve of 41 thermal decomposition reaction mechanism functions was carried out by general integral method, such as General Integral equation [17], Maccallum Tanner equation [18] and Agrawal equation [19], and the most available mechanism functions and the corresponding thermal decomposition kinetic equation were obtained. The reliability of the processing method is judged by comparing the consistency of the processing results of Kissinger's method and Ozawa's method, etc.

The most probable kinetic model function is shown in Equation 4:

$$G(\alpha) = (1-\alpha)[-\ln(1-\alpha)]^{\frac{2}{3}} \tag{4}$$

The final reaction mechanism of the integral form, differential form and the corresponding thermal decomposition kinetic equation are shown below:

$$\frac{d\alpha}{dt} = 3 \times 10^{8.64287} exp\left(-\frac{97.3765}{RT}\right)(1-\alpha)[-\ln(1-\alpha)]^{\frac{2}{3}} \tag{5}$$

$$[-\ln(1-\alpha)]^{\frac{1}{3}} = 10^{8.64287} exp\left(-\frac{97.3765}{RT}\right) t \tag{6}$$

3.3 Thermal safety assessment

For energetic materials, T_{SADT} (self-accelerating decomposition temperature) and T_{bpo} (critical temperature of thermal explosion) are two important characteristic parameters, which are used to ensure thermal safety and thus ensure their storage and operation safety.

T_{SADT} and T_{bpo} can be calculated by Equation 7 and Equation 8 respectively.

$$T_{bpo} = \frac{E_K - \sqrt{E_K^2 - 4E_K R T_{po}}}{2R} \tag{7}$$

$$T_{SADT} = T_{bpo} - \frac{R T_{bpo}^2}{E_K} \tag{8}$$

For CA-NW, Tbpo =476.6K, and TSADT=456.6K.

4. Conclusions

CA-NW was a new primary explosive with high orientation. Its thermal decomposition behavior was studied to evaluate its thermal safety. The thermal decomposition reaction was observed by DSC and TG-DTG methods.

(1) The thermal decomposition of CA-NW presents one intensive exothermic process. The kinetic equation of the decomposition process is

$$\frac{d\alpha}{dt} = \frac{3 \times 10^{8.64287}}{\beta} \exp\left(-\frac{97.3765}{RT}\right)(1-\alpha)[-\ln(1-\alpha)]^{\frac{2}{3}}$$

(2) T_{SADT} and T_{bpo} are 456.6K and 476.6K, respectively.

(3) Results show that the thermal stability of CA-NW is stable at normal temperature and doesn't decompose.

References

[1] *The encyclopedia of explosives and related items* vol 1 p 532-534
[2] Forhar F *et al* 2008 The adsorption of hydrazoic acid on single-walled carbon nanotubes *Mater. Chem. Phys.* vol 112 p 427-431
[3] Pelletier V et al 2010 Copper Azide Confined Inside Templated Carbon Nanotubes *Adv. Funct. Mater.* vol 20 p 3168-3174
[4] Currano L *et al* 2010 Energetic materials for integration on chip *14th International Detonation Symposium* p 163-169
[5] Guo R *et al* 2012 A micro initiator realized by integrating KNO3@CNTs nanoenergetic materials with a Cu microbridge *Chemical Engineering Journal* vol 211 p 31-36
[6] Jean D 2009 MEMS micro-detonator based fuzing *The 53rd Annual Fuze Conference*
[7] Cope R 2007 NDIA's 51st annual fuze conference UNITED STATES navy overview *The 51st Annual Fuze Conference*
[8] Mitchell S 2006 NDIA's 50th annual fuze conference UNITED STATES navy overview *The 50th Annual Fuze Conference*
[9] Zhang F *et al* 2012 Preparation and characterization of copper azide nanowire array *Mater. Letters* vol 89 p 176-179
[10] Wang Y *et al* 2016 Detonation transfer performance of detonation wave driven-Ti flyer by copper azide *Journal of Beijing Institute of Technology* vol 25(suppl.1) p 216-220
[11] Zhang F *et al* 2016 Integrated microheater-copper azide chip *Journal of Beijing Institute of Technology* vol 25(suppl.1) p 210–215
[12] Zhang F *et al* 2012 In-situ Preparation of a Porous Copper Based Nano-Energetic Composite and Its Electrical Ignition Properties *Propellants, Explosives, Pyrotechnics* vol 38 p 41-47
[13] Wang Y *et al* 2016 In-situ Preparation of Carbon Nanotubes Array Confined with Copper Azide *Chinese Journal of Energetic Materials* vol 24(4) p 386-392(in Chinese)
[14] Zhang Z *et al* 2015 In-situ Preparation and Characterization of Porous Copper Azide. *Initlators & Pyrotechnics* vol 2 p 26-28(in Chinese)
[15] Kissinger H 1957 Reaction kinetics in differential thermal analysis *Analytical Chemistry* vol 29 (11) p 1702-1706
[16] Ozawa T 1965 A method of analyzing thermogravimetric data. *Bulltin of Chemical Society Jpn*, vol 38(11) p 1881-1886
[17] Coats A *et al* 1964 Kinetic Parameters from Thermogravimetric Data *Nature* vol 201(4914) p 68 – 69
[18] MacCallum J *et al* 1970 The kinetics of thermogravimetry *Eur. Polymer J.* vol 6(7) p 1033-1039
[19] Agrawal R 1987 A new equation for modeling nonisothermal reaction *J. Therm. Anal.* vol 32(1) p 149-156
[20] Hu R *et al* 2008 *Thermal Analysis Kinetics (2th)* Beijing Science Press (in Chinese)

Preparation and Characterization of Direct Write Explosive Ink Based on CL-20

Lei Zhang[1,2,*], Fang Zhang[2], Yanlan Wang[2], Ruishan Han[2] and Li Yang[1]

[1] State Key Laboratory of Explosion Science and Technology, Beijing Institute of Technology, Beijing 100081, China
[2] Science and Technology on Applied Physical Chemistry Laboratory, Shaanxi Applied Physics-chemistry Research Institute, Xi'an 710061, Shaanxi, China

E-mail:ggy628@163.com

Abstract. In this paper, a new CL-20-based explosive ink formulation used for direct written was designed and prepared. The explosive ink can be smoothly written by microinjection and precisely patterned explosive film can be obtained. The CL-20 explosive ink was characterized by SEM and XRD, and DSC was used to analysis the thermal decomposition performance of the explosive ink. The critical thickness of the CL-20 explosive ink was 0.153mm and the average detonation velocity was 8088.9m/s.

Keywords: explosive ink,CL-20, direct write technology, detonation performance

1. Intorduction

With the development of micro electro-mechanical system (MEMS) fuzing technology, it requires that energetic materials must have micro- or nano-scale to ensure their activity [1,2]. In addition, the charging method for MEMS fusing must have features such as high precision, automation, intelligence, and safety. There are two traditional ways of explosive charging: mold pressing and cast molding [3-6]. The method of mold press molding is relatively simple, but the shape of the energetic agent is limited by the mold, and this method is mostly used for large charging. The cast molding method can produce uniform charging and the combustion stability is good. However, the solidification of the drug column is slow, and when casting a small-sized explosive column, the surface tension and the viscosity affect the molding accuracy, which cannot meet the charging demand for MEMS fusing.

Direct-write technologies provide a high-throughput processing platform capable of precision patterning. The direct write technology of energetic materials is to mix the energetic material with binder in a certain ratio and formulate the energetic ink material with fluidity[7,8]. The computerized motion control program can be used to complete the explosive writing. However, the performance of the direct write explosive inks is influenced by very complicated process parameters due to the challenging requirements of the structures. Except for satisfying the conventional requirements such as stability, viscosity, surface tension, and uniformity, the ink must also have a property that a considerable loading density can be achieved after drying.

Currently, 2,4,6,8,10,12-hexanitrohexaazaisowurtzitane (HNIW, CL-20) with high energy density (about 2.0 g/cm), excellent detonation pressure (about 44.8 GPa) and detonation velocity (about

9762m/s), which is regarded as one of the most promising explosive to replace HMX and RDX [9-11]. In this paper, we chose CL-20 as the energetic material to prepare a direct write explosive ink, which can be smoothly written and have high energy density and micro-scale dimensions, and could be used in MEMS. Then, the explosive ink samples were characterized by scanning electron microscope (SEM), X-ray diffraction(XRD) and differential scanning calorimeter (DSC). Finally, the written critical size and the detonation velocity were also measured, which could demonstrate the performance of this explosive ink.

2. Experiments

2.1. Instruments and Reagents

TESCAN VEGA TS5136XM Scanning Electron Microscope(Czech), OXFORD INCA 300 X-ray Energy Dispersive Spectrometer, BRUKER D8 advance X-ray Diffraction(Germany), NETZSCH DSC204F1 Differential Scanning Calorimeter(Germany), SHOTMASTER-300Ωx Three Dimensional Mobile Platform(Japan), DL32/16 velocity measuring instrument (China).

ε-CL-20, purity greater than 99.5%, provided by North University of China; Polyvinyl alcohol (Tianjin Ke Miou Chemical Reagent Co., Ltd.); Ethyl cellulose (Shanghai Jing Chun Biochemical Technology Co., Ltd.); Isopropyl alcohol (Tianjin Chemical Reagent Factory). All chemicals used in this investigation were of analytic grade.

2.2. Preparation of explosive ink

The direct write explosive ink used in this paper was composed of nano ε-CL-20, binder system and other additives. The nano ε-CL-20 was characterized with SEM (Fig.1a), and the results show that the crystal size of nano ε-CL-20 was about 200~300nm. The binder system includes the binder and the solvent. In this paper, two different binder systems were designed with polyvinyl alcohol and ethyl cellulose, and the solvent was H_2O and isopropyl alcohol. The mixture was stirred until the solids has been completely dissolved. The dry ε-CL-20 was added to the solution and then the explosive ink was obtained evenly after stirring. The total solids compositions in the ink is about 60%, and the composition ratio of ε-CL-20 to the binder is 19:1. A commercial inkjet printer (SHOTMASTER-300Ωx Three Dimensional Mobile Platform, Japan) was used to print and pattern explosive materials. With this device, the explosive ink can be written directly onto a surface or deposited into a cavity, and the writing parameters can be precisely controlled. The direct write explosive ink shown in figure 1 b.

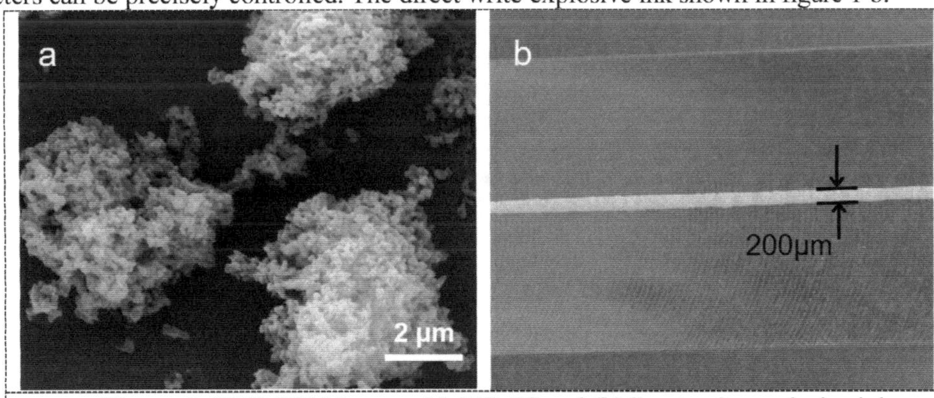

Figure 1. XRD SEM image of **(a)**CL-20 and **(b)**direct write explosive ink.

3. Characteristics and Discussions

3.1. X-Ray Diffraction Analysis

X-ray diffraction was used to characterize the crystal structure and the results are shown in figure 2. Comparing the X-ray diffraction curves of ε-CL-20 (Figure. 2a) and CL-20 slurry (Figure. 2b) we can see that the strong diffraction peaks of the Bragg angles of CL-20 used for explosive ink is at 2θ = 12.580°, 13.830° and 30.310°, respectively, which are same with the PDF#00-050-2045 (RDB) characteristic peaks of ε-CL-20 [12]. When made into ink slurry, the crystal shape of CL-20 remains in the ε-phase, which indicates that the configured ink solution does not change the crystalline form of ε-CL-20, so its detonation performance can still be well maintained.

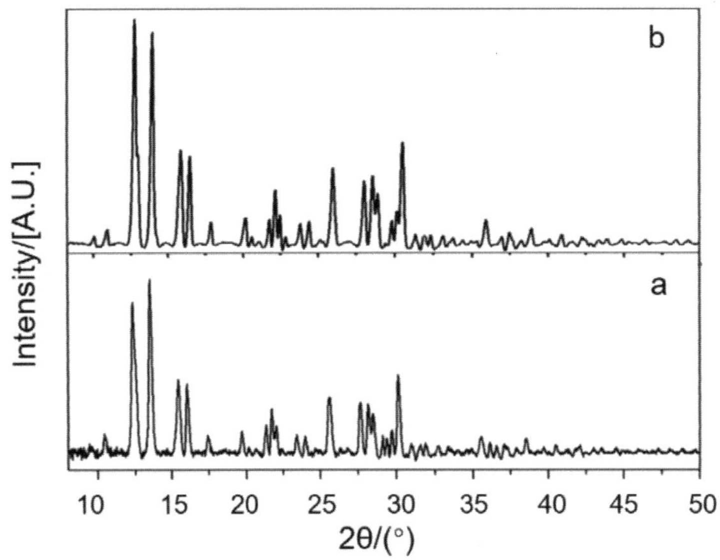

Figure 2. XRD patterns of (**a**)CL-20 used for slurry, (**b**)direct write explosive ink after mixture.

3.2. Thermal Properties

In order to evaluate the thermal properties, we characterized the ε-CL-20 and direct write explosive ink with the NETZSCH DSC204F1 Differential Scanning Calorimeter (sample quality: about 1.0 mg, reference sample: α-Al_2O_3, atmosphere: flowing nitrogen, average heating rate: 10 °C /min, temperature range: 25 °C to 500 °C).The results is shown in Figure 3.

It can be observed that the curve of ε-CL-20 rises more abruptly with a peak temperature at 242.04°C, which was approached to the peak temperature at 249.4 °C observed by Dong [13] et al. And the second curve of the direct write explosive ink based on CL-20 has its own thermal decomposition temperature at 239.62°C.

Through the comparison n of the figure 3(a)(b), it can be seen that, compared with the raw ε-CL-20, the peak temperature of the explosive ink has no obvious change, which indicates that the bond system is not involved in the thermal decomposition process.

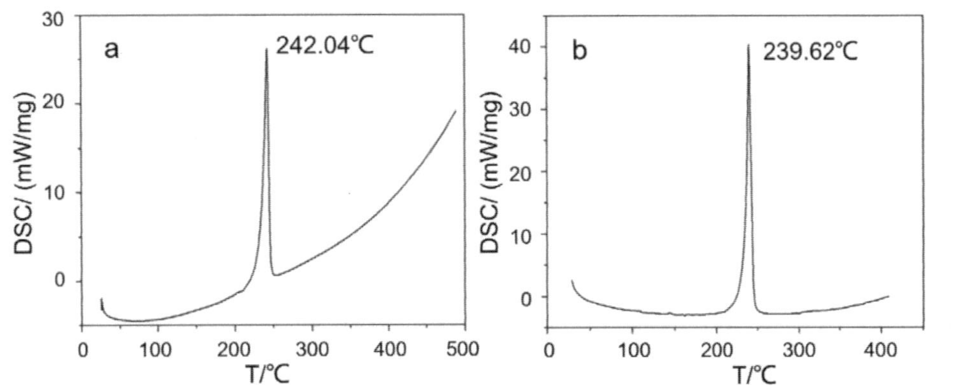

Figure 3. DSC curve of **(a)**ε-CL-20 and **(b)** the explosive ink obtained in a temperature range from 25 to 500 °C with an average heating rate of 10 °C /min.

3.3. The Critical Diameter of detonation

Every explosive has its critical diameter of detonation which indicates that below which the detonation cannot be maintained. For this explosive inks, the critical size is related to the explosive components and the density of the explosive material within the ink. In order to get a minimum and accurate critical size, two standard steel rules(the thickness is1mm and the length is 150mm) were fixed on a aluminum block with varying distance from 0 to 1mm, which constitute a shim stock to direct write the explosive ink, after dried and initiated with an LD-14detonator. In order to more effectively evaluate the critical size performance of energetic inks, we prepared three specimens(1#, 2# and 3#), which were averaged to determine their critical dimensions.The results are shown in figure 4. Through the formula of D=(A/L)·ΔL we can get the critical size of this explosive ink. The parameters of explosive ink and the experiment results of the average critical size are shown in table 1.

Table 1. The parameters of average detonation velocity.

No.	Density L/(mm)	Diameter A/mm	Distance ΔL/ mm	Critical size D/mm	Average velocity Đ/(mm)
1#	10	2300	3499	4738	8074.6
2#	10	1947	3208	4450	8024.1
3#	20	2780	5252	7766	8019.6

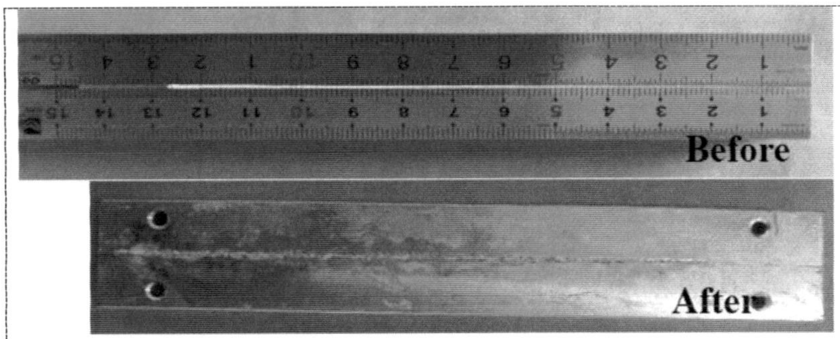

Figure 4. The result of critical diameter of detonation testing.

3.4. Average detonation velocity

Detonation velocity is the main parameter for characterizing the detonation performance of explosives. In this paper, the average detonation velocity of ε-CL-20 explosive ink is obtained using a DL32/16 time interval measuring instrument, which is an 8-channel, Chinese-manufactured. The range of time interval between each channel is from 50ns~820ms, the time resolution is 1ns, which can improve the accuracy and reliability of the test results effectively. We place the enamelled copper wires at the bottom of the explosive ink as shorting pins. Making and breaking of shorting pins produces a pulse signal to measuring instrument, and then the interval time between each channel would be recorded. Using those time values, the average detonation velocity can be calculated. The schematic diagram of detonation velocity measured by enamelled copper weirs method is shown in figure 5.

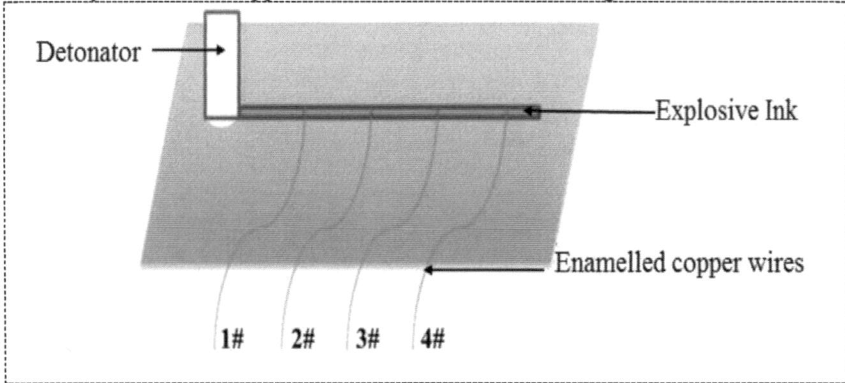

Figure 5. Schematic diagram of detonation velocity measured by enameled copper weirs method.

To ensure the accuracy of the data obtained, we set the enamelled wires with an average of 10mm and 20mm, those distances are measured accurately with the vernier caliper to reduce errors. The experiment results of the average detonation velocity are shown in table 2. The four target lines used in this experiment were disconnected at the beginning, and when the detonation wave passed, they turned on due to ionization. The time interval Δt for detonation waves passing through the two target lines is calculated by Δt=t(i+1)-ti: The interval between each two targets is d. Calculated by the formula v=d/Δt, the average velocity v between detonation waves passing through two target lines can be obtained. It can be seen that under the same distance condition of four enamelled wires, the maximum velocity measured was 8237.1 m/s, and the average detonation velocity of those four test specimens were 8088.9 m/s. This value is close to the detonation velocity of CL-20's press charging (9061m/s,1.932g/cm) [14], which means that this ink formulation improves CL-20's charging process without affecting its detonation performance.

Table 2. The parameters of average detonation velocity.

No.	Distances mm	Channel 1 T/ns	Channel 2 T/ns	Channel 3 T/ns	Channel 4 T/ns	Average velocity V/(m/s)
1#	10	2300	3499	4738	6018	8074.6
2#	10	1947	3208	4450	5686	8024.1
3#	20	2284	4676	7086	9570	8237.1
4#	20	2780	5252	7766	10262	8019.6

4. Conclusions

This study demonstrates the preparation of direct write explosive ink based on ε-CL-20, the results of XRD and DSC show that the crystal type of ε-CL-20 is not changed in the explosive ink slurry and its thermal decomposition temperature is at 239.62°C, which is approached to the peak temperature of pure ε-CL-20.Moreover, through the critical size test at different diameters when the CL-20 solid content of direct write explosive ink is 60%, it can propagate the detonation waves successfully in the groove at 0.134 mm, and the largest detonation velocity of this explosive ink can reach 8237.1m/s. The above performances make the explosive ink a promising aditive to be used in MEMs explosive train.

References

[1] Chopin hua *et al* 2012 Low-Cost MEMS Initiators *MicroAssembly Technologies* inc
[2] Oliva M Leonardo *et al* 2005 Spectroscopic Characterization of Thermal Inkjet Transfer of Deposits of Energetic Materials *Third LACCET* Cartagena de Indias Colombia
[3] Andrew Ihnen *et al* 2010 Inkjet Printing of Nanocomposite High-Explosive Materials for Direct Write Fuzing *54th Annual Fuze Conferenee* Kansas NV
[4] K.K.B. Hon *et al* 2008 Direct writing technology-Advances and developments. C*IRP Annals - Manufacturing Technology* Vol 57 p 601-620
[5] Pique Alberto and Douglas B 2002 Direct-Write Technologies for Rapid Prototyping Applications *Academic press* San Diego Ca
[6] Le HP 1998 Progress and Trends in Ink-Jet Printing Technology *Journal of Imaging Science and Technology* Vol 42(1) p 49-62
[7] J. L. Zunino *et al* 2010 Inkjet Printed Devices for Armament Applications *NSTI-Nanoteeh* Vol 2 p 542-545
[8] Eric Windso 2010 Application of Inkjet Printing Technology to Produce Test Materials of 1,3,5-Trinitro-1,3,5 Triazcyclohexane for Trace Explosive Analysis *Analytical Chemistry* Vol 82(20) p 8519-8524
[9] Monteil-Rivera F *et al* 2004 Physico-chemical measurements of CL-20 for environmental applications *Journal of Chromatography A* Vol 1025(1) p 125-132
[10] Brain E F *et al* 2010 Development, performance and use of direct write explosive inks. *14th International Detonation Symposium* Coeur d'Alene Idaho
[11] Shen Longsheng *et al* 2015 Formulation Design of Energetic Ink and Its Application in Microporous Charge. *Explosive Materials* Vol 5(44) p 1-5
[12] Song Zhenwei *et al* 2010 Crystal Transition of ε-CL-20 in Different Solvent *Chinese Journal of Energetic Materials* Vol 6(18) p 648-653
[13] Dong Haishan *et al* 2002 ThermoGRAMS of Energetic Materials *National Defense Industry Press*
[14] NAN Yuxiang *et al* 2015 JWL Equation of state of Detonation Product for CL-20 Based Pressed Composite Explosive *Chinese Journal of Energetic Materials* Vol 6(23) p 516-521

Study on the Design of bridge microstructure for Energy Conversion Components

REN Wei[1, 2], ZHANG Bin[2], CHU Enyi[2], ZHAO Yulong[1], YIN Ming[2], LI Hui[2], WANG Kexuan[2]

1. The State Key Laboratory for Manufacturing Systems Engineering, Xi'an Jiaotong University, Xi'an, 710049, China;
2. Science and Technology on Applied Physical Chemistry Laboratory, Shaanxi Applied Physics-Chemistry Research Institute, Xi'an, 710061, China

E-mail address: rw0192@163.com (Ren Wei)

Abstract: Microstructure energy conversion components, as the key components of MEMS initiating explosive device, has a significant impact on the output performance and energy utilization of the initiating explosive device, especially the design of its bridge microstructure. In order to perfect the design theory of microstructure energy conversion components of MEMS (Micro Electro-Mechanical Systems) initiating explosive device, 8 different bridge microstructure are designed for energy conversion components in this paper, and using simulation research and infrared test to carry out the research on the planar microstructure effect and the better bridge type choice of a series of bridge type. Obtained two optimization bridge structure of microstructure energy conversion components and the influence law between different microstructure and output performance of energy conversion components, and the average ignition voltage of V-50 bridge microstructure energy conversion components up to 100μF/3.5V, and the energy utilization rate is 46.6%。

Key Words: microstructure energy conversion component; bridge structure; temperature rise rate; bridge fusing current

1 Introduction

Initiating explosive device (IED) is the disposable components with detonation, ignition and powered, which has the characteristic of function first, sensitive, widely used and etc. With the research and development of Micro-Electro-Mechanical-Systems (MEMS), MEMS initiating explosive device became a research hotspot in the field of IED in recent years[1]. MEMS initiating explosive device which base on the MEMS technology or compatible with MEMS process, it has some size characteristics: the structure of energy conversion components and the scale of energy materials in micro, the scale of core device in the sub-millimeter level, and the scale of system in millimeter level. This kind of initiating explosive device is integrated with information control, safe and actuator and detonating unit, bring it has the characteristics of information, small and multi-function, and it's a key basic technology of informatization and smart ammunition[2-4].

Microstructure Energy Conversion Components is the core devices of IED to realize energy conversion, and it generally is a plane resistance that made of metal film or semi-conducting material in substrate, and to realize the energy transfer and implementation with energetic materials through resistance heating and so on. The focus in the study of MEMS IED is the study on microstructure energy conversion components, and research has been widely spread in several research institutes at home and abroad, which focus on substrate materials[5,6], the structure of bridge[7-9] and the matching with sequence energy, etc. Research indicates that lower ignition energy and the high energy conversion efficiency are the main characteristics of ideal energy conversion components of MEMS IED[10].

At present, the research of microstructure energy conversion components are relatively more, but have not formed a series of study. For example, the influence of bridge structure on performance of energy conversion components, which most study is aimed at a single bridge structure, and did not comprehensive comparison of the performance for different bridge structure energy conversion components. Based on this, eight different bridge structures energy conversion components were designed in this paper, and to develop research of performance for them, and obtained a better bridge structure and the influence law of different bridge structures for energy conversion components performance.

2 Design research for bridge of microstructure energy conversion components

2.1 Bridge design of microstructure energy conversion components

At present, bridge design of microstructure energy conversion components mainly includes the square, inverted "V" shape, serpentine, etc, and considering resistivity scale effect of metal film at the same time, and eight different bridge structures energy conversion components were designed, that are shown in Figure 1. Bridge material of energy conversion components is Ni-Cr alloy, and the substrate is Pyrex7740 glass.

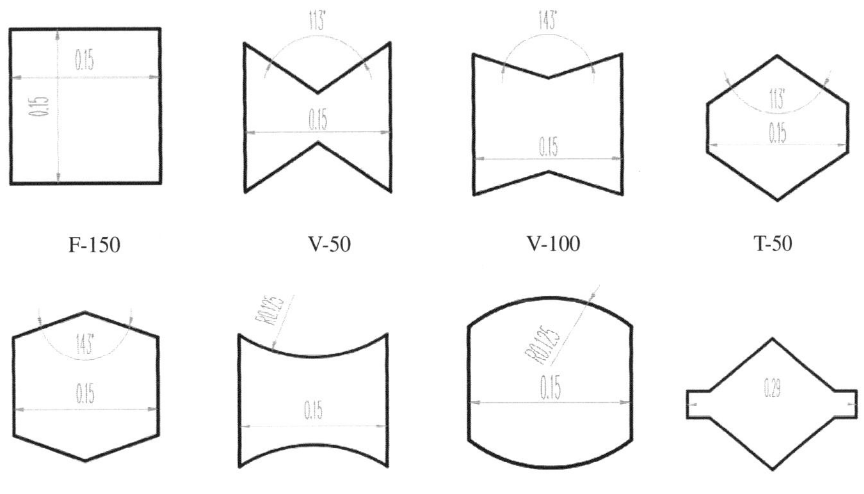

T-100 S-100 S-150 L-1

Figure 1. Design structure of bridge for energy conversion components（units: mm）

The above design bridge structure is based a square bridge (150μm*150μm，F-150) to transform. The inverted "V" structure, which bridge width on both sides to keep 150 micron, and the center of the bridge area is reduced to 50 micron and 100 micron respectively, and to obtain two different inverted "V" bridge structure, and named V-50 and V-100. The trapezoidal structure, which bridge center width keep 150 micron, and the bridge width on both sides is reduced to 50 micron and 100 micron respectively, and to obtain two different trapezoidal bridge structure, and named T-50 and T-100. The curve structure, which bridge width on both sides to keep 150 micron, and the center of the bridge area is reduced to 100 micron, and to obtain a hyperbolic bridge structure, and named S-100; And the bridge center width to keep 150 micron, and the bridge width on both sides is reduced to 100 micron, and to obtain an ellipse bridge structure. The diamond structure, which bridge center width keep 150 micron, and the diamond sides are 150 micron, and the diamond on each side extension of 15 microns to make two electrode distance increased to 290 micron, and named L-1.

2.2 Simulation research for microstructure energy conversion components

Initial bridge area of MEMS energy conversion components is solid, and the temperature in the bridge area increases gradually with the increase of the energy conversion time. When the temperature in the bridge area is over bridge material melting point, and the bridge area gradually change from solid to liquid, which flows, namely bridge fusing. Thus, for the different bridge structure energy conversion components, the simulation analysis of electro-thermal and fluid-solid heat transfer is mainly carried out, and to obtain the temperature rise curves under different input conditions, highest temperature in bridge area and the distribution of fluid-solid heat transfer region. Then, the optimal bridge structure can be obtained by comparative analysis.

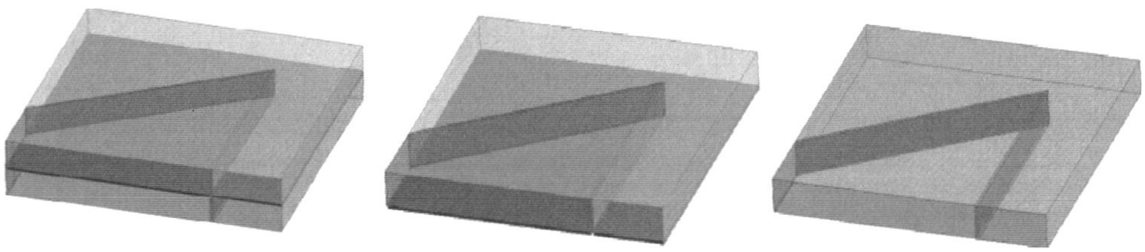

Population geometric model Bridge area geometric model Basal geometric model

Figure 2. 1/4 geometric models of microstructure energy conversion components

In the process of simulation research, substrate and outside air domain such as the influence of the interface must be considered, except for the energy conversion components model. Therefore, the exchange of heat with the outside domain after heating, melting and solidification of material, and the air natural convection are main consideration in the simulation calculation. Due to the energy conversion components have symmetry plane structure, and the study was carried out mainly based on as shown in Figure 2 1/4 simulation of geometric model, and to obtain the temperature rise rate and the temperature distribution in different bridge structure under the condition of constant current, and comparison of fluid-solid heat exchange area.

IOP Conf. Series: Journal of Physics: Conf. Series **1209** (2019) 012017 doi:10.1088/1742-6596/1209/1/012017

Simulation calculation using two kinds of incentives, including constant current and capacitor discharge. Constant current excitation is from 100 mA to 600 mA, and gradient 100 mA, and capacitance incentive is 100μF/3V. Material of energy conversion layer is Ni-Cr alloy, and substrate is Pyrex7740 glass, and their thermo-dynamic parameters are shown in Table 1.

Table 1. Thermodynamic parameters of two kinds of materials

materials	density (kg/m^3)	melting point (K)	specific heat (J/kg K)	thermal conductivity (W/m·K)
Pyrex7740	2500	1525	753	1.4
Ni-Cr	8400	1673	460	15

The curve of temperature rise rate that simulation calculation in different bridge structure of energy conversion components under the condition of constant current excitation (200mA) is shown in Figure 3.

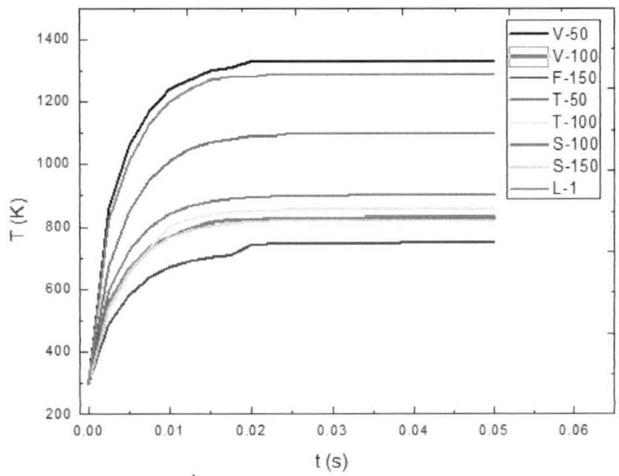

Figure 3. Temperature rise rate under constant current excitation of energy conversion components with different bridge structure (200mA)

The figure shows that temperature rise rate of energy conversion components with different bridge structure have obvious difference in 200 mA constant current excitation, and the temperature rise rate of V-50 and L-1 is significantly higher than other bridge, and the output temperature is relatively high.

Then, comparative analysis fluid solid heat transfer area of energy conversion components with different bridge structure, without considering obvious defects that the manufacture craft process on the microstructure of bridge caused, which can effectively predict the broken bridge position after ignition for energy conversion components with different bridge structure, and to guide the design of bridge microstructure. Fluid solid heat transfer area distribution of eight kinds of energy conversion components shown in Figure 4.

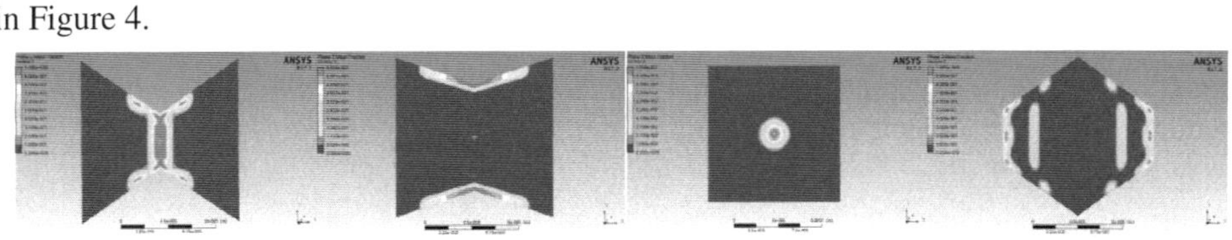

V-50 V-100 F-150 T-50

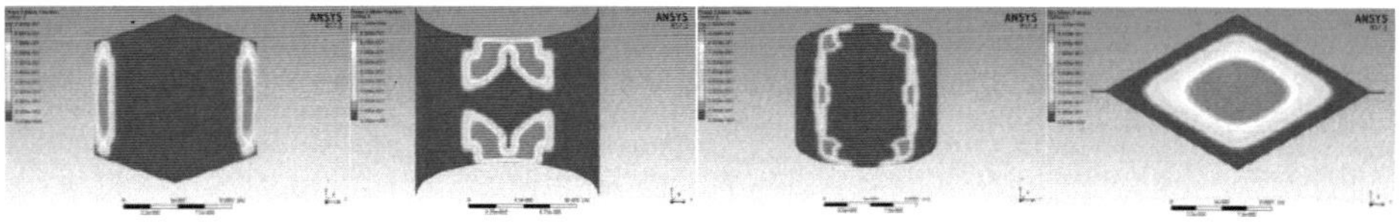

T-100 S-100 S-150 L-1

Figure 4. Fluid-solid heat transfer of microstructure energy conversion components in different bridge

The figure shows that other bridge structures give priority to fluid-solid heat exchange at the corner except for the F-150, S-150 and L-1 bridge structure, and where the heat is concentrated that is the location of the most vulnerable bridge. Among them, the V - 50 bridge as an example, the center of the bridge is the narrowest, and the current density is the maximum during the energizing process, and heat generated much and relatively concentrated, therefore, it is the easiest to break the bridge. Fluid-solid heat transfer of F-150 bridge occurred preferentially at the center of the bridge, that because the bridge structure is simple without local structure, and the heat is evenly distributed throughout the bridge, and the superposition of heat lead to the highest temperature in the center of the bridge area. Fluid-solid heat transfer of L-1 bridge occurred at the center of the bridge, rather than on both sides, and the analysis shows that the line width is related to the 15 microns length of both sides of the original design. Fluid-solid heat transfer of S-150 bridge occurred neither on either side of the bridge nor in the bridge center, which is caused by the characteristics of the bridge microstructure. The heat concentrated generated on up and down the center of the S-150 bridge, and heat accumulation leads to the highest temperature in the center of the bridge area with the increase of current time energy conversion components.

In conclusion, through the simulation and analysis research for eight kinds of energy conversion components, not only can effectively predict the broken bridge position after ignition, more important is can according to the ignition performance requirements, and further improving the design structure of MEMS energy conversion components.

2.3 Infrared test for microstructure energy conversion components

In order to get in response characteristics of energy conversion components under the constant current, and validate the simulation results, infrared test was performed for eight kinds of energy conversion components under the constant current excitation. And the temperature rise curve, critical fuse current in bridge area and the broken bridge location were obtained by infrared test, and further clear the influence law of bridge microstructure for response output of energy conversion components under constant current. Temperature rise curve of infrared test for energy conversion components with different design structure under constant current excitation (200mA) is shown in Figure 5.

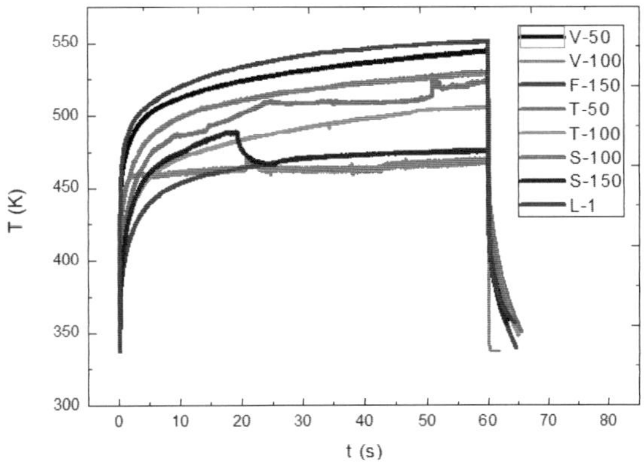

Figure 5. Temperature rise under constant current excitation of energy conversion components with different bridge structure (200mA)

The figure shows that the simulation calculation results and the infrared test results are basically identical, the temperature rise rate of V-50 bridge and L-1 bridge is significantly higher than other bridge, and the output temperature is relatively high. The analysis shows that there is an obvious inflection point in the bridge area, and the current density in the inflection point is bigger, therefore, the temperature rise of these type energy conversion components is higher than others.

Critical fuse current in bridge area and the corresponding output temperature of energy conversion components with different microstructure are shows in Tab.2.

Table 2. Bridge fusing current of microstructure energy conversion components with different bridge structure

Bridge	V-50	V-100	F-150	T-50	T-100	S-100	S-150	L-1
bridge fusing current/mA	250	300	325	225	250	225	300	225
output temperature/K	595	547	616	569	594	616	631	645

3 Ignition performance test for microstructure energy conversion components

Capacitive ignition performance experiment was carried out for eight kinds of energy conversion components, and the ignition circuit as shown in Figure 6, and the capacitance is 100 uF. And test at same time using oscilloscope gathering the curve of current, voltage change over time, and experiment was carried out in isolation room.

Test samples are shown in Figure 7, the chip of energy conversion components were paste to the PCB board, and using ultrasonic welding to realize the conduction between the chip of energy conversion components and PCB board, and it is convenient to apply electric excitation to carry out the ignition performance test. And the energetic materials are lead styphnate.

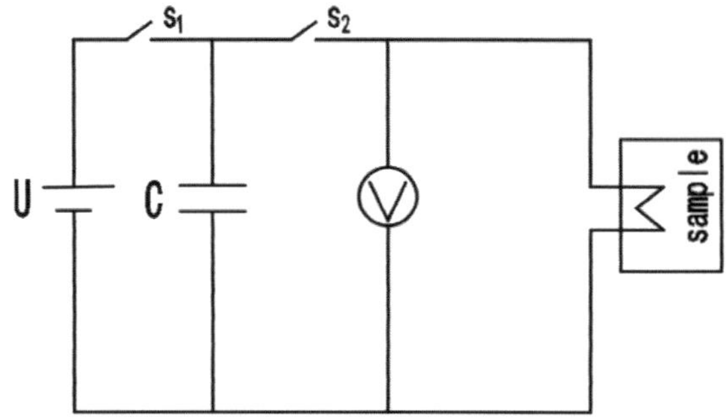

Figure 6. Firing circuit diagram of microstructure energy conversion components

Figure 7. The test samples of microstructure energy conversion components

Through simple processing by the curve of the voltage over time, the *P(t)*—*t* was obtained and shown in Figure 8. Then, to integral curve to obtained the bridge input energy (*Q*), and to calculate the power input energy (*E*) according to equation (2), finally, to calculate the energy utilization rate (η) of energy conversion components according to equation (3).

$$Q = P(t) \cdot t = I^2(t) \cdot R_0 \cdot t \quad (1)$$

$$E = \frac{1}{2} C U^2 \quad (2)$$

$$\eta = \frac{Q}{E} \quad (3)$$

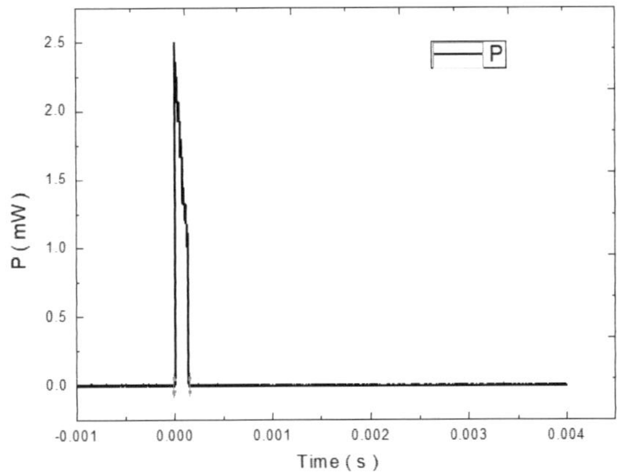

Figure 8. The curves between *P(t)* and *t* of microstructure energy conversion components

Ignition performance parameters that obtained by ignition test for energy conversion components with different bridge are shown in Tab.3.

Table 3. the firing performance parameters of microstructure energy conversion components with different (capacitance 100μF)

bridge	V-50	V-100	F-150	T-50	T-100	S-100	S-150	L-1
test resistance/Ω	6.57	5.57	5.23	8.65	6.58	9.17	8.14	8.66
firing voltage/V	3.5	4.5	5.5	5	5	5	6	4.5
ignition time/μs	138	160	143	150	139	198	176	181
E/mJ	0.61	1.01	1.51	1.25	1.25	1.25	1.8	1.01
Q/mJ	0.286	0.367	0.601	0.555	0.353	0.548	0.779	0.334
η*100	46.6	36.3	39.7	44.4	28.2	43.9	43.3	33

The ignition performance test results shows that:

(1) Among the eight kinds of microstructure energy conversion components, the energy utilization of V-50 bridge structure is the highest, reached 46.6%, and the ignition voltage, ignition energy and ignition time are less than other bridges.

(2) When the resistance is less than a certain critical value (10Ω), the influence of the resistance for ignition voltage is irregular. The analysis shows that the small resistance (≯10Ω), the ignition performance of microstructure energy conversion components were mainly impacted by design microstructure, the quality and the heat dissipation area of bridge.

(3) Ignition test results of energy conversion components with different bridge microstructure and simulation research, and the infrared test results are consistent, and that is the ignition performance of V-50 bridge structure is the best.

4 Conclusion and perspectives

Microstructure energy conversion components is the key components of MEMS initiating explosive device to realize energy conversion, and the bridge structure of which has a significant impact on the output performance and energy utilization of the initiating explosive device. Through the simulation research, and the infrared test and ignition test for eight kinds of microstructure energy conversion components, the optimization bridge structure has been obtained and the influence law that bridge microstructure for ignition performance.

(1) Different design microstructure of energy conversion components have a significant impact on the output performance, and the "neck" structure is helpful to improve the ignition performance, that reflected in the V-50 bridge structure energy conversion components has the highest output temperature, and the fastest temperature rise rate; In addition, fluid-solid heat exchange priority occurred at the corner of bridge area, and the bridge area where centralized heat is the most easily broken bridge location.

(2) When the resistance is less than a certain critical value (10Ω), the influence of the resistance for ignition voltage is irregular. The analysis shows that the small resistance ($\ngtr 10\Omega$), the ignition performance of microstructure energy conversion components were mainly impacted by design microstructure, the quality and the heat dissipation area of bridge.

Subsequently, to further study on low-energy for microstructure energy conversion components, and to study on the information integration technology, so as to promote and expand the application research for microstructure energy conversion components.

References

[1] ZHANG Bin, CHU En-yi, REN Wei, et al. Research Progress in Energy Conversion Components for MEMS Initiating Explosive Device[J]. Chinese Journal of Energetic Materials, 2017, 25(5):428-436.

[2] YANG Wen-mao, LIU Yan-wen, XU Lu-xiang, et al. Review of Film Growth by Sputtering Technology[J]. Journal of Vacuum Science and Technology, 2005, 25(3): 204-210.

[3] KONG Jun-feng, LI Bing. A new generation of initiation technique and application [J]. Technology Foundation of National Defence, 2010(7):40-43.

[4] SHEN Rui-qi YE Ying-hua DAI Shi-zhi. Conception and Application of Digital Initiation Technique[J]. Initiators & Pyrotechnics, 2000(2):37-39.

[5] Zhang K L, Chou S K, Ang S S. Fabrication, modeling and testing of a thin film Au/Ti microheater[J]. International Journal of Thermal Sciences, 2007, 45(6):580-588.

[6] Chopin Hua. Low-cost MEMS initiators[C]//MicroAssembly Technologies[C]//NDIA 54th Annual Fuze Conference, 2010.

[7] Neha Mehta, TACOM-ARDEC, Chemical Engineer, Design and development of micro energetic initiators (MEI)[C]//NDIA 47th Annual Fuze Conference, 2003.

[8] XIE Rui-zhen, REN Xiao-ming, WANG Ke-xuan, et al. Effect of bridge parameters on the firing sensitivity of Ni-Cr alloy thin film transducer elements[J]. Initiators & Pyrotechnics, 2012(1):18-20.

[9] Lee J, Kim T. MEMS solid propellant thruster array with micro membrane igniter[J]. Sensors & Actuators A Physical, 2013, 190(2):52-60.

[10] HU Yan, YE Ying-hua, SHEN Rui-qi, et al. Preparation and Characterization of Al-CuO Reactive Bridge Film[J]. Chinese Journal of Energetic Materials, 2010, 18(3):339-342.

IOP Conf. Series: Journal of Physics: Conf. Series **1209** (2019) 012018 doi:10.1088/1742-6596/1209/1/012018

Roll angle estimation of guided gun-launched projectile using MEMS angular rate gyro and magnetometer

R Z Qian[1,2],G T Xu[1,2],K H Yang[1,2] and Y Y Xu[1,2]

[1]Science and Technology on Electromechanical Dynamic Control Laboratory, 710065 Xi'an China

[2]Xi'an Institute of Electromechanical Information Technology, Xi'an 710065, China

giggsqrz@163.com

Abstract. The previous method to estimate roll angle of guided gun-launched projectile can be completely invalid in case of GPS is disabled in battlefield. To solve that problem, this paper proposes roll angle estimation used MEMS angular rate gyro and magnetometer. Comparing to the previous method with GPS, the addition of magnetometer outputs increases the measurement accuracy, but increases the dimension and nonlinearity of the measurement model. Therefore the Extended Kalman Filter is introduced. Through trajectory simulation, this study indicates that the absolute roll estimation error is less than 7 degrees after 3 seconds in whole flight trajectory, which is completely met the guided system requirement.

1. Introduction

Facing the harsh requirement of lower cost and higher precision in modern warfare, Conventional shells urgently need to improve the accuracy of the second dimension, in which high-precision rolling angle measurement is critical to the accuracy of the whole guiding system.

In previous methods, Global Positioning System (GPS) and angular rate are combined in roll angle estimation[1]. Artillery shell's roll angle is related to pitch rate, yaw rate and gyroscope's output in Body Frame coordinate, pitch rate and yaw rate can calculated from velocity information of GPS. However GPS may be disabled under the certain circumstance in the battlefield, there has to be an immune method to solve the roll angle measurement anyway.

In case GPS signal is not available, a magnetometer can be used as an important measurement[2]. It is assumed that variation of earth magnetism can be ignored when the artillery projectile's flight distance is shorter than 100km, in another words, earth magnetism is constant in the study. Three-axis magnetometer is used to measure the geomagnetic field signal in the projectile coordinate system, considering the nonlinearity of the measurement model, combining with Extended Kalman Filter (EKF) to measure.

2. Previous Estimation Using Satellite Position System and Angular Rate Gyro [3~5]

This section introduces the previous roll estimation method, which uses the satellite position information and angular rate gyro outputs.

Figure 1 shows a Body Frame (b) and Navigation Frame (n) for a gun-launched projectile. The x axis of nb is defined by a forward direction of the projectile, the y axis is defined by an up direction perpendicular to the x axis and the z axis is defined by a right direction perpendicular to the x-y plane, while $x_n y_n z_n$ is defined as Navigation Frame. A pitch rate ($\dot{\theta}$) and a yaw rate ($\dot{\psi}$) defined in Body Fixed Plane and a roll defined in Body Frame. The attitude of gyroscope is aligned with the Body Frame.

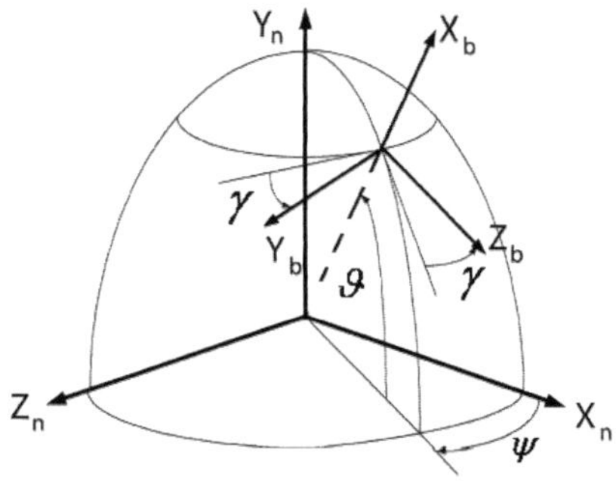

Figure 1. Body Frame and Navigation Frame.

In figure 1, it is assumed that guided projectile rotates with respect to x axis by roll angle (γ), so the y axis and z axis outputs of gyroscope will measure projection of the pitch rate and the yaw rate with respect to roll angle's sinusoidal variation when the shell is flying. According to this knowledge, following equation is given.

$$\begin{cases} \omega_y^b = \dot{\theta}\cos\gamma + \dot{\psi}\sin\gamma \\ \omega_z^b = -\dot{\theta}\sin\gamma + \dot{\psi}\cos\gamma \end{cases} \tag{1}$$

The pitch rate ($\dot{\theta}$) and yaw rate ($\dot{\psi}$) mentioned above are obtained from the GPS velocity as shown in equation (2) and equation (3).

$$\theta = \tan^{-1}\left(\frac{-V_y}{\sqrt{V_x^2 + V_z^2}}\right) \tag{2}$$

$$\psi = \tan^{-1}\left(\frac{V_y}{V_x}\right) \tag{3}$$

Then the pitch rate ($\dot{\theta}$) and yaw rate ($\dot{\psi}$) are held by equation (4) and equation (5) as follow.

$$\dot{\theta} = \frac{\Delta\theta}{t} \tag{4}$$

$$\dot{\psi} = \frac{\Delta\psi}{t} \tag{5}$$

Combining equation (4) and equation (5), the analytic formula of gyroscope's output in Body Frame and pitch/yaw rate in Body Fixed Plane are established, and roll angle calculation is available.

3. Roll Angle Estimation Using Gyro and Magnetometer without GPS

In this section, a method of using gyro and magnetometer without GPS's data is introduced, then an Extended Kalman Filter is also brought in because of the high non-linear measurement model.

3.1. Coordinate Definition[3]

Normally, earth magnetism consists of two parts: declination and inclination. Magnetic declination is defined as an angle between the magnetic north and the geographic north. Magnetic inclination is defined as an angle between a magnetic vector and a tangential plane of the earth. It is also assumed that both declination and inclination are constant because trajectory of gun-launched shell is mostly nearer than 100km.

An attitude of a magnetometer is exactly aligned with a Body Frame (x_b, y_b, z_b). Each axis of magnetometer output is same to earth magnetism in the Body Frame. Equation (6) presents the coordinate transformation and the matrix C_n^b is coordinate transform matrix from a Navigation Frame to a Body Frame.

$$\begin{bmatrix} x_b \\ y_b \\ z_b \end{bmatrix} = C_x^3 C_z^2 C_y^1 \begin{bmatrix} x_n \\ y_n \\ z_n \end{bmatrix} = C_n^b \begin{bmatrix} x_n \\ y_n \\ z_n \end{bmatrix} \tag{6}$$

Where $C_y^1 = \begin{bmatrix} \cos\psi & 0 & -\sin\psi \\ 0 & 1 & 0 \\ \sin\psi & 0 & \cos\psi \end{bmatrix}, C_z^2 = \begin{bmatrix} \cos\theta & \sin\theta & 0 \\ -\sin\theta & \cos\theta & 0 \\ 0 & 0 & 1 \end{bmatrix}, C_x^3 = \begin{bmatrix} 1 & 0 & 0 \\ 0 & \cos\gamma & \sin\gamma \\ 0 & -\sin\gamma & \cos\gamma \end{bmatrix}$.

When the earth magnetic frame (x_{em}, y_{em}, z_{em}) is introduced, the y and z axes are determined by the coordinate transformation from the Navigation Frame to the Earth Magnetic Frame. In this coordinate transformation, a yaw is declination (β), a pitch is inclination (α) and a roll is zero. The normalized magnetometer output in Earth Magnetic Frame is $[1 \quad 0 \quad 0]^T$, which is a theoretical value. Equation (7) represents coordinate transformation from the Navigation Frame to the Earth Magnetic Frame.

$$\begin{bmatrix} 1 \\ 0 \\ 0 \end{bmatrix} = \begin{bmatrix} \cos\beta & \sin\beta & 0 \\ -\sin\beta & \cos\beta & 0 \\ 0 & 0 & 1 \end{bmatrix} \begin{bmatrix} \cos(-\alpha) & 0 & -\sin(-\alpha) \\ 0 & 1 & 0 \\ \sin(-\alpha) & 0 & \cos(-\alpha) \end{bmatrix} \begin{bmatrix} x_n \\ y_n \\ z_n \end{bmatrix} = C_n^{em} \begin{bmatrix} x_n \\ y_n \\ z_n \end{bmatrix} \tag{7}$$

Thus the earth magnetometer's output in Body Frame is as equation (8) as follow.

$$\begin{bmatrix} x_e^b \\ y_e^b \\ z_e^b \end{bmatrix} = C_n^b C_{em}^n \begin{bmatrix} x_{em} \\ y_{em} \\ z_{em} \end{bmatrix} = C_n^b (C_{em}^n)^{-1} \begin{bmatrix} 1 \\ 0 \\ 0 \end{bmatrix} \tag{8}$$

3.2. System Model and Measurement Model for Magnetic Measurement Using EKF

A state vector, a measurement vector and a system model are shown below.

$$x_k = [p \quad \gamma_0 \quad \omega_y \quad \omega_z \quad \psi \quad \theta]^T \tag{9}$$

$$z_k = [\sin\gamma(k) \quad \cos\gamma(k) \quad \dot{\psi} \quad \dot{\theta} \quad x_m^b \quad y_m^b \quad z_m^b]^T \tag{10}$$

$$x_{k+1} = H_{6\times6} \times x_k + q_k \tag{11}$$

Where $\gamma(k)$ is roll angle and q_k is system noise. Then measurement matrix for EKF is given following.

$$H_1 = \begin{bmatrix} k\cos(pk+\gamma_0) & \cos(pt+\gamma_0) & \dfrac{\theta}{\dot{\theta}^2+\dot{\psi}^2} & -\dfrac{\dot{\psi}}{\dot{\theta}^2+\dot{\psi}^2} \\ -k\sin(pk+\gamma_0) & -\sin(pk+\gamma_0) & \dfrac{\dot{\psi}}{\dot{\theta}^2+\dot{\psi}^2} & \dfrac{\dot{\theta}}{\dot{\theta}^2+\dot{\psi}^2} \\ 0 & 0 & \cos(pk+\gamma_0)/\cos\alpha & -\sin(pk+\gamma_0)/\cos\alpha \\ 0 & 0 & \sin(pk+\gamma_0) & \cos(pk+\gamma_0) \end{bmatrix}$$

$$H_2 = \frac{\partial\left(\begin{bmatrix} x_e & y_e & z_e \end{bmatrix}\right)^T}{\partial p} \quad H_3 = \frac{\partial\left(\begin{bmatrix} x_e & y_e & z_e \end{bmatrix}\right)^T}{\partial \gamma} \quad H_4 = \frac{\partial\left(\begin{bmatrix} x_e & y_e & z_e \end{bmatrix}\right)^T}{\partial \psi} \quad H_5 = \frac{\partial\left(\begin{bmatrix} x_e & y_e & z_e \end{bmatrix}\right)^T}{\partial \theta}$$

$$H_k = \begin{bmatrix} & H_1 & & 0_{4\times 2} & \\ H_2 & H_3 & 0_{3\times 2} & H_4 & H_5 \end{bmatrix} \tag{12}$$

4. Simulation Results

In this section, an algorithm based on angular rate gyro and magnetometer using EKF without GPS velocity is simulated. Table 1 shows simulation conditions.

Table 1. Simulation Conditions.

Characteristics	Value
Roll rate	10～60 rad/s
Initial roll	0～360 deg (random)
Initial pitch	70 deg
Initial yaw	0 deg
Algorithm step	1 ms
Gyro bias	200 deg/hr
Rate noise density	0.3 deg/sec/hr$^{0.5}$
Non-orthogonal angle	3 deg
Magnetometer noise	2×10^{-3}

Figure 2 shows estimation error is nearly zero using theoretical analystic formula without measurement noise that indicates algorithm's validity. In figure 3, however, estimation error is drastically increase because of measurement noise, then an EKF is needed to bring into algorithm instead of analysis formula.

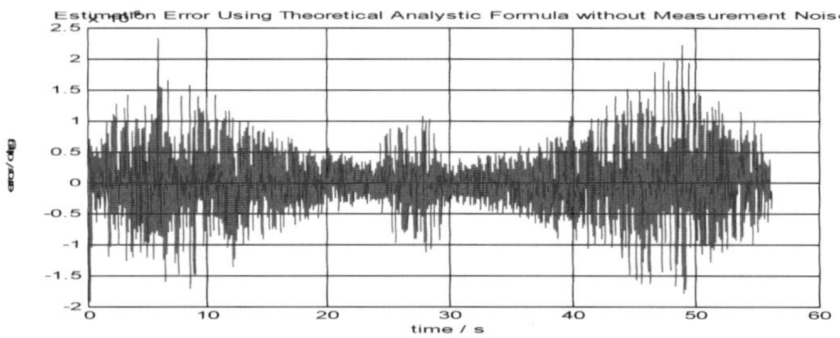

Figure 2. Estimation without Measurement Noise.

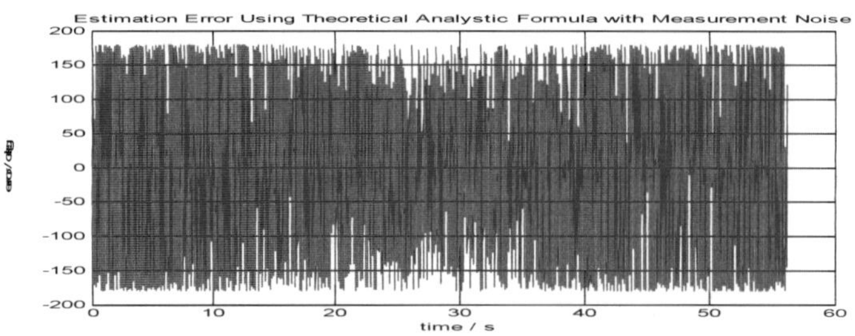

Figure 3. Estimation with Measurement Noise.

Figure 4 shows real roll angle and calculated roll angle using EKF, figure 5 indicates estimation error using EKF.

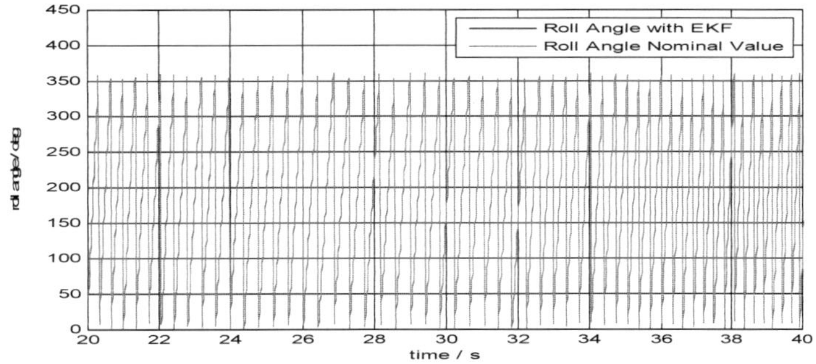

Figure 4. Roll Angle with EKF and Roll Angle Nominal Value.

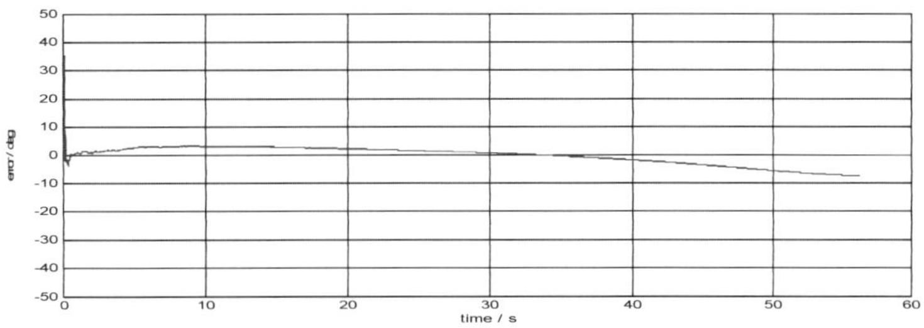

Figure 5. Estimation Error Using EKF.

In figure 5, the absolute roll estimation error is less than 7 degrees after 3 seconds in whole flight trajectory, which is fully met the guided system requirement. The reason of larger estimation error in the beginning and the final trajectory is inaccurate measurement noise this simulation chooses, estimation precision can be improved by enhancing the gyro and magnetometer measurement model.

5. Conclusion

This paper discusses the previous roll estimation method, then introduces a new way to estimate the roll angle using gyro and magnetometer without GPS and an Extended Kalman Filter algorithm is also brought in to analyze. The simulation results show that the absolute roll estimation error is less than 7 degrees after 3 seconds in whole flight trajectory, which is completely met the guided system requirement.

References

[1] Lucia D J 1995 *Estimation of the Local Vertical State for a Guided Munition Shell with an Embedded GPS/Micro-Mechanical Inertial Navigation System*(MS Thesis MIT Cambridge Messachusetts)

[2] Hee Young Park Kwnag Jin Kim Jang Gyu Lee and Chan Gook Park June 2007 *Roll Angle Estimation for Smart Munitions* (France: IFAC Symposium on Automatic Control in Aerospace, Toulouse) P 25-29

[3] Han Sung Lee Kwang Jin Kim Hee Young Park Chan Gook Park and Jang Gyu Lee August 2008 *Roll Estimation of a Smart Munition Using a Magnetometer Based on an Unsented Kalman Filter*(AIAA Guidance, Navigation and Control Conference and Exhibit Honolulu Hawaii) P18-21

[4] Zhou Xiang Huo Pengfei and Qi Keyu 2010 *Attitude Measurement of Gun-Launched Projectile Based on Three-Axis Magnetometer and MEMS Angular Rate Gyroscopes* (Journal of Detection & Control)

[5] S. Haykin 2001 *Kalman Filtering and Neural Networks* (New York:Jony Wiley & Sons, Inc.)

Acknowledgments

IOP Conf. Series: Journal of Physics: Conf. Series **1209** (2019) 012018 doi:10.1088/1742-6596/1209/1/012018

This study is accomplished with a fundamental research (614260102030217) of Key State Laboratories. The author wishes to express sincere gratitude to all fellows working together in Science and Technology on Electromechanical Dynamic Control Laboratory.

IOP Conf. Series: Journal of Physics: Conf. Series **1209** (2019) 012019 doi:10.1088/1742-6596/1209/1/012019

Minimum damping profile of micro/nano-robot and as the carrier for drug delivery: theory study

L X Yang[1], G C Zhang[1], N Fan[1], J Guo[1,3] and B Peng[1,2 *]

[1] Department of Mechanical and Electrical Engineering, University of Electronic Science and Technology of China, Chengdu, Sichuan, P.R. China
[2] Center for Robotics, University of Electronic Science and Technology of China, Chengdu, Sichuan, P.R. China
[3] Department of Materials Science and Engineering, Xiangtan University, Xiangtan, Hunan, P.R. China
[*]Corresponding author: beipeng@uestc.edu.cn

Abstract. The use of the micro/nano-robot as a carrier for drug delivery in vivo represent a novel and innovative method in cancer/tumor therapy and related biomedical applications. In this paper, the authors firstly deduce the minimum damping profile of micro/nano-robot as it locomoting in vascular and capillaries system. Where originating from the Bernouli's equation, and integrating the calculus of variations with the Euler-Lagrange equation, then the profile of micro/nano-robot which the generatrix of this axisymmetric body is the 4/3 order of the curve is obtained. Following, the elastic responses of the cell indented by this profile are studied detailedly. Utilizing the method of continuum mechanics, the interior stress and displacement distributions, indentation depth, surface stress distribution (corresponding the critical rupture stress of cytomembrane) and applied force (corresponding the required force output from micro/nano-robot) are derived. Furthermore, those elastic responses indented by conical and spherical profile are computed and contrasted. Conclusively, the authors provide a design reference for the profile of micro/nano-robot as a carrier for drug delivery, and deduce the elastic responses of the cell when it penetrated by the body of micro/nano-robot.

1. Introduction

The single-cell-based nanotechnologies have been very popular in the cancer/tumor therapy [1], induced pluripotent stem cells (iPs) technology [2], medical imaging technology and related cell-engineering fields, which require delivering the nanoscale objects (therapeutic drug, or the special biomolecular factors/gene, or the medical imaging material) into the cell. The easy implement, high precision and effective target transfection are key considerations for evaluating the way of this transporting. Summarily, there are two main patterns (active and passive pattern) to complete this transportation in vivo/vitro for the single cell.

Electroporation is a typical active pattern of transportation, which combines the microfluidic with microchannel technology, and actuated by microelectrode [3], making up a nanofountain-probe (NFP) that can deliver the nanoscale objects into the cell in vitro [4]. High transportation efficiency can be achieved through electroporation, yet the damage of cell by electric field also can't be ignored. In addition, those methods are not facilitate to implement in vivo.

Endocytosis is the other main way but a passive pattern, which utilize the natural properties of cell membrane-endocytosis effect and mediated by nanoparticle [5], to complete the drug delivery/transportation [6]. Yet the diverse and intricate biochemical environment and fluctuating force of cell membrane, will give rise to (may misleading the bonding) low efficiency binding between nanoparticle and targeted diseased cell membrane.

Based on the analysis above, all limitations of the ways of electroporation and endocytosis will be overcame if there will be a smart micro/nano-robot [7], that be capable of initiative self-locomoting in vascular and actively penetrating the cytomembrane for drug delivery in vivo [8]. For this newly arisen robotics research area there are many related works have been done [9]. Especially for the propulsion principle which is crucial component for the safe and sustainable operation in vivo, including magnetic, ultrasound, optical, thermal, electrical, acoustic, and enzyme-functionalized, have been put forward for the micro/nano-robot [10]. However, the quantitative evaluation of penetrating force output from the micro/nano-robot which also is the applied force to puncture the cytomembrane have not been investigated, yet will be investigated in current paper.

In this paper, the authors firstly deduce the minimum damping profile of micro/nano-robot as it locomoting in vascular and capillaries system. Where originating from the Bernouli's equation, and utilizing method of calculus of variations and Euler-Lagrange equation, then profile of micro/nano-robot which the generatrix of this axisymmetric body is the 4/3 order of the curve is obtained. Secondly, the elastic responses of the cell penetrated by this profile of micro/nano-robot are studied systematically. Based on the elementary solution of axisymmetric problem in elastic mechanics and employing the method of contact mechanics, the interior stress and displacement distributions, the indentation depth, the normal surface stress distribution (provide a quantitative indicator to understand the critical rupture stress of cell membranes), the applied load (corresponding the required force that micro/nano-robot shall output), and their interrelationships are investigated. Finally, cases of conical and spherical profile of micro/nano-robot also be calculated and contrasted.

2. Minimum damping profile of micro/nano-robot

As the micro/nano-robot locomoting in the blood capillary or extracellular matrix, its geometric shape which suffering minimal resistance is required that within a certain propulsion power. Nonetheless, the microfluidic effect have not been considered that may make important influence in the moving process as micro/nano-robot locomoting in microfluidic environment. Yet regarding the microenvironment as the incompressible flow and leaving out the friction effect between the interface, therefore the Bernouli's equation [11] was introduced as follow

$$\frac{v^2}{2}+g\cdot z+\frac{p}{\rho}=C \tag{1}$$

Moreover, the gravity effect can be neglected as the negligible difference at height dimension. Taking those factors into consideration, and consider the pressure along the x-coodinate that the micro/nano-robot suffering, as shown in figure 1-(a) which the micro/nano-robot locomoting in vascular and capillary medium.

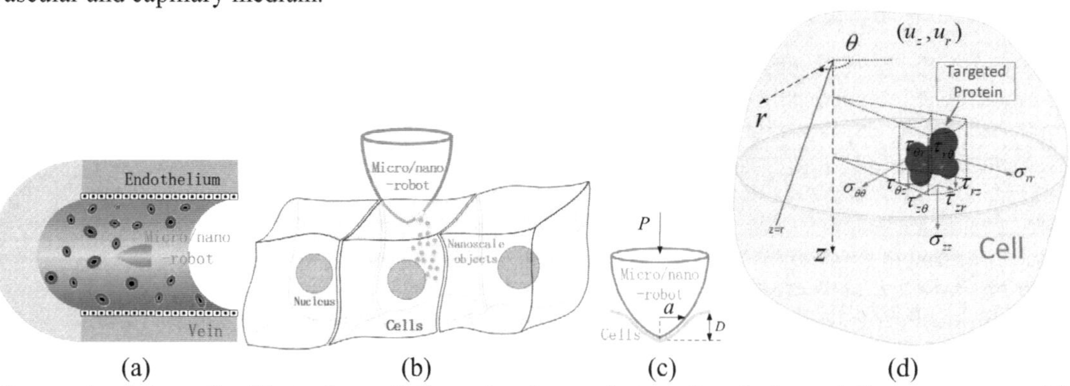

Figure 1. Schematic illustration of the micro/nanorobot-mediated drug delivery process. (a). Schematic of the micro/nano-robot locomoting in vascular and capillaries system, where the

minimum damping profile of micro/nano-robot is required. (b). When the maicro/nano-robot arriving at the area of diseased cells, it will going on to actively puncture the cell membrane to inject nanoscale objections into the cell. (c). In the indentation process there are four significant variables needed to be considered: the surface stress distribution on the cytomembrane, the applied load *P*, and the contact radius a and the penetrating depth *D*, which can also be used in related field like measurement of modulus of softmatter. (d).The accompanying stimulation of interior stress and displacement on the certain(targeted) protein inside the cell also are important factors for evaluating this whole processes.

Then Bernouli's as above can be reduced and transformed to,

$$p_x = c_1 \cdot \rho \cdot v^2 \cdot \sin^2 \theta \tag{2}$$

where p_x is the components of pressure along the *x*-axis at chosen point c_1 is the arbitrary constant ρ is the density of the fluid at all points in the fluid, v is the speed of fluid flow at a point on a streamline.

Following, the per unit force of micro/nano-robot received can be obtained by multiply the pressure at a point with the unit area of thrust surface. Combining with the simple mathematical derivation the per unit force can be got

$$dF_x = 2\pi \cdot c_1 \cdot \rho \cdot v^2 \cdot y \cdot \left(y'\right)^3 \cdot dx \tag{3}$$

Further, the functional expression of the total force that the micro/nano-robot received under it's a certain profile of head shape $y(x)$, noted with $J[\square]$, that can be formulated as

$$J\left[y(x)\right] = \int_F dF_x = 2\pi \cdot c_1 \cdot \rho \cdot v^2 \cdot \int_0^L \cdot y \cdot \left(y'\right)^3 \cdot dx \tag{4}$$

Among above expression, introducing the notation $W = y \cdot \left(y'\right)^3$ and considering the one-dimensional Euler-Lagrange equation [12], which given by

$$\frac{\partial W}{\partial y} - \frac{d}{dx}\frac{\partial W}{\partial y'} = 0 \tag{5}$$

and incorporating with the boundary conditions

$$y(0) = 0, y(L) = R \tag{6}$$

where L is the length and R is the radius at length of L of the revolved axisymmetric profile of body of micro/nano-robot. Then the profile of minimum damping force that the micro/nano-robot suffering when it locomoting in vascular systems, can be deduced as

$$y = L^{-3/4} \cdot R \cdot x^{3/4} \tag{7}$$

For keeping accordance with following processing in cylindrical coordinates, the function of the micro/nano-robot profile can be written as follow,

$$z = L \cdot R^{-4/3} \cdot r^{4/3} \tag{8}$$

3. Elastic responses of cell indented by the micro/nano-robot

With the above derived profile of micro/nano-robot from fluid mechanics, then it will going on to actively puncture the cytomembrane to inject the nanoscale obicetions into the cell, as shown in figure 1(b). The quantitative understanding of the normal surface stress distribution on cytomembrane indented by this profile, the corresponding required output force from micro/nano-robot and the interior displacement/stress distribution will be studied, as shown in figure 1(c-d). Different from adopting means of the molecular dynamics which considering the interaction potential between two atoms (or particles), but the elastic mechanics especially contact mechanics was employed to study this penetrating process.

3.1 Interior stress and displacement components

IOP Conf. Series: Journal of Physics: Conf. Series **1209** (2019) 012019 doi:10.1088/1742-6596/1209/1/012019

Regarding the whole problem is axisymmetric and the cylindrical coordinates system is adopted where the displacement components are (u_r, u_z) and the stress components are ($\tau_{zr}, \sigma_{zz}, \sigma_{rr}, \sigma_{\theta\theta}$) while other components are vanish. The Young modulus of the micro/nano-robot was assumed to be stiffer far more than the cell's modulus and the target cell can be approximately treated as an isotropic semi-space. There are two boundary conditions in this problem, as follow

$$u_z\left(0<r<a, z=0\right)=D-g\left(r\right), \qquad \sigma_{zz}\left(r\geq a, z=0\right)=0 \tag{9,10}$$

where a is the radius of the contact area, D is arbitrarily parameter which regarded as the depth of the penetrating, $g(r)$ is formulated to characterize the profile features of the micro/nano-robot.

Based on the elementary solutions deduced by Sneddon [13], the components of displacements and stress variables can be expressed in form of function $\psi(\zeta)$ and considering $r=\rho\cdot a$, thus

$$u_z(\rho,z)=H_0\left[\psi(\zeta)\cdot\left(\frac{\lambda+\mu}{\lambda+2\mu}\cdot\frac{z}{a}+\frac{1}{\zeta}\right)\cdot\zeta^{-1}\cdot e^{-\frac{\zeta}{a}z};\zeta\to\rho\right] \tag{11}$$

$$\sigma_{zz}(\rho,z)=-\frac{2\mu\cdot(\lambda+\mu)}{\lambda+2\mu}\cdot H_0\left[\psi(\zeta)\cdot\left(1+\frac{\zeta}{a}\cdot z\right)\cdot\frac{1}{\zeta\cdot a}\cdot e^{-\frac{\zeta}{a}z};\zeta\to\rho\right] \tag{12}$$

where H_n denotes the Hankel transform, λ and μ are known as Lame constants, and the $\psi(\zeta)$ denotes the function that related the profile of micro/nano-robot.

Combining with the boundary conditions, the above two solutions can be reduced to a pair of dual integral equations which had been solved by Sneddon[13]. Substituting the minimum damping profile of micro/nano-robot (as shown in Equation (8) into the solution of dual integral, thus

$$\psi(\zeta)=\frac{2}{\pi}\cdot\left\{D\cdot\sin\zeta-\frac{L}{2}\cdot\left(\frac{a}{R}\right)^{4/3}\cdot\beta\left(\frac{5}{3},\frac{1}{2}\right)\cdot\zeta\cdot\left[\cos\zeta+\zeta\cdot\int_0^1 t^{7/3}\cdot\sin(\zeta\cdot t)\cdot dt\right]\right\} \tag{13}$$

and substituting the $\psi(\zeta)$ into the normal stress components Equation (12) and considering the stress on the contact periphery $\sigma_{zz}(r=a, z=0)=0$, then having

$$2D-\frac{7}{3}\cdot L\cdot\left(\frac{a}{R}\right)^{4/3}\cdot\beta\left(\frac{5}{3},\frac{1}{2}\right)=0,\quad \psi(\zeta)=\frac{28}{9\pi}\cdot L\cdot\left(\frac{a}{R}\right)^{4/3}\cdot\beta\left(\frac{5}{3},\frac{1}{2}\right)\cdot\int_0^1 t^{1/3}\cdot\sin(\zeta\cdot t)\cdot dt \tag{14,15}$$

Furthermore, other four components which firstly derived by Sneddon also can be rewrote in form of function $\psi(\zeta)$:

$$u_\rho(\rho,z)=H_1\left[\psi(\zeta)\cdot\left(\frac{\lambda+\mu}{\lambda+2\mu}\cdot\frac{\zeta}{a}\cdot z-\frac{\mu}{\lambda+2\mu}\right)\cdot\zeta^{-2}\cdot e^{-\frac{\zeta}{a}z};\zeta\to\rho\right] \tag{16}$$

$$\tau_{z\rho}(\rho,z)=-\frac{2\mu\cdot(\lambda+\mu)}{\lambda+2\mu}\cdot H_1\left[\psi(\zeta)\cdot\frac{z}{a^2}\cdot e^{-\frac{\zeta}{a}z};\zeta\to\rho\right] \tag{17}$$

$$\begin{aligned}\sigma_{\rho\rho}(\rho,z)=&-\frac{2\mu\cdot(\lambda+\mu)}{\lambda+2\mu}\cdot H_0\left[\psi(\zeta)\cdot\left(1-\frac{\zeta}{a}\cdot z\right)\cdot e^{-\frac{\zeta}{a}z}\cdot\frac{1}{\zeta\cdot a};\zeta\to\rho\right]\\&+\frac{2(\lambda+\mu)\cdot\mu}{\rho\cdot a\cdot(\lambda+2\mu)}\cdot H_1\left[\psi(\zeta)\cdot\left(\frac{\mu}{\lambda+\mu}-\frac{\zeta}{a}\cdot z\right)\cdot\zeta^{-2}\cdot e^{-\frac{\zeta}{a}z};\zeta\to\rho\right]\end{aligned} \tag{18}$$

$$\sigma_{\theta\theta}(\rho,z) = -\frac{2\mu\cdot\lambda}{\lambda+2\mu}\cdot H_0\left[\psi(\zeta)\cdot e^{-\frac{\zeta}{a}z}\cdot\frac{1}{\zeta\cdot a}\cdot J_0(\zeta\rho);\zeta\to\rho\right] - \frac{2(\lambda+\mu)\cdot\mu}{\rho\cdot a\cdot(\lambda+2\mu)}\cdot H_1\left[\psi(\zeta)\cdot\left(\frac{\mu}{\lambda+\mu}-\frac{\zeta}{a}\cdot z\right)\cdot\zeta^{-2}\cdot e^{-\frac{\zeta}{a}z};\zeta\to\rho\right] \tag{19}$$

By substituting the expression of $\psi(\zeta)$ as shown in Equation (15) into the Equations (11,12,16-19), all elastic responses of the displacement and stress components in the cell which indented by the minimum damping profile of micro/nano-robot $z = L\cdot R^{-4/3}\cdot r^{4/3}$ can be obtained respectively.

3.2 Penetration depth, normal surface stress and applied force

Noting the result about penetrating depth in Equation (14), it can be easily calculated as

$$D = \frac{7}{6}\cdot L\cdot\left(\frac{a}{R}\right)^{4/3}\cdot\beta\left(\frac{5}{3},\frac{1}{2}\right) \tag{20}$$

Furthermore, by substituting Equation (15) into Equation (12) and setting z=0, the surface normal stress can be computed as

$$\sigma_{zz}(r,0) = \frac{-E}{(1-v^2)}\cdot\frac{14}{3\pi}\cdot\beta\left(\frac{5}{3},\frac{1}{2}\right)\cdot\frac{L}{R}\cdot\left(\frac{a}{R}\right)^{1/3}\cdot\left\{{}_2F_1[-\frac{1}{6},\frac{1}{2},\frac{5}{6},\left(\frac{r}{a}\right)^2]-\left(\frac{r}{a}\right)^{1/3}\cdot\sqrt{\pi}\cdot\Gamma(\frac{5}{6})\cdot\Gamma^{-1}(\frac{1}{3})\right\} \tag{21}$$

Finally, the applied force P is defined as

$$P = -2\pi\cdot\int_0^a \sigma_{zz}(r,0)\cdot r\cdot dr \quad 0<r<a \tag{22}$$

by substituting Equation (21) into above expression and integrating with Equation (20), the applied force P depends on penetrating depth D can be evaluated as

$$P = \frac{E}{(1-v^2)}\cdot\frac{8}{7}\cdot\left[\frac{7}{6}\cdot\beta\left(\frac{5}{3},\frac{1}{2}\right)\right]^{-3/4}\cdot L^{-3/4}\cdot R\cdot D^{7/4} \tag{23}$$

Actually, the same results of penetration depth D, normal surface stress distribution and applied force P also can be calculated from known formulas [14], and the difference from here is the mathematical derivations where the obtain of the formula of normal surface stress distribution is based on the full known of $\sigma_{zz}(r,z)$ and then setting z=0, whereas the interior stress distributions are failed to be presented in existing classical formulas.

4. Comparison of elastic responses indented by three different profiles

In this section, the values of interior stress and displacement distributions, penetrating depth, normal surface stress, and applied force indented by three different profiles of micro/nano-robot will be computed and contrasted successively. Where the generatrix of the axisymmetric indenter of the micro/nanorobot: 1 order curve of the generatrix (conical profile), 4/3 order curve of the generatrix (minimum damping force), 2 order curve of the generatrix (spherical profile), will be studied respectively.

4.1 Comparison of interior stress and displacement components

Like the treatment methods in Section 3 where the obtainment of function of $\psi(\zeta)$ for the minimum damping profile (4/3 order curve of the generatrix of indenter), the functions of $\psi^{Conical}(\zeta)$ and $\psi^{Spherical}(\zeta)$ for conical and spherical profile can be derived analogously as below,

$$\psi^{Conical}(\zeta) = k\cdot a\cdot\frac{1-\mathrm{Cos}\,\zeta}{\zeta}, \quad \psi^{Spherical}(\zeta) = \frac{4}{\pi}\cdot\frac{a^2}{R}\cdot\left(\frac{\sin\zeta}{\zeta^2}-\frac{\cos\zeta}{\zeta}\right) \tag{24,25}$$

where k is the slope of conical indenter, R is the radius of spherical indenter.

By substituting the functions of $\psi(\zeta)$, $\psi^{Conical}(\zeta)$ and $\psi^{Spherical}(\zeta)$ into the Equations (11-12,16-19) respectively, the corresponding stresses and displacements distributions indented by those three different profiles of micro/nano-robot can be obtained. For the less computation and the convenience of comparison, the values of stress and displacement only on the line 1 (*z=r*) in the interior of cell (half space), as depicted in figure 1-(d), are numerically calculated as shown in figure 2.

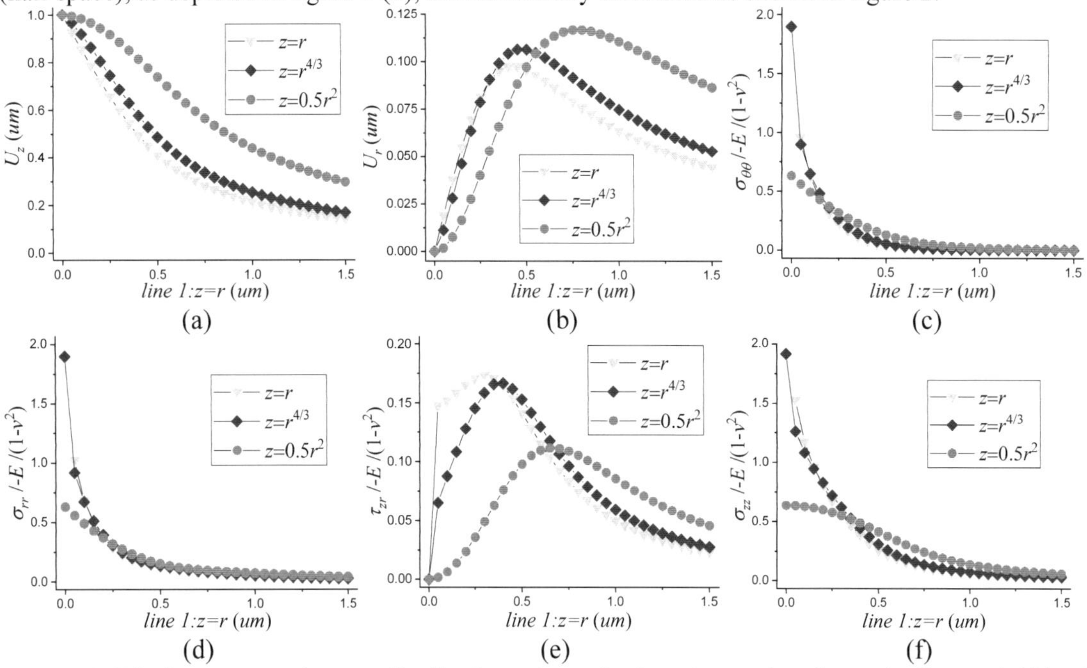

Figure 2. Displacement and stress distributions along the *line 1: z=r* (as shown in figure. 1-(d)) of half space (cell) indented by three different profiles of maicro/nano-robot, where the generatrix of axisymmetric indenter is $z=r$; $z=r^{4/3}$; $z=0.5r^2$ respectively. And the range of *line 1* ($z=r$: $0\sim1.5$ *(um)* with interval distance 0.05 *(um)*), the Poisson ratio of half space (cell) $v=0.49$ and the unified penetration depth $D=1$ *(um)* are set. (a) and (b) indicate the values of two displacement component u_z and u_r. (c)-(f) show the values of four stress components σ_{rr}, τ_{zr}, $\sigma_{\theta\theta}$ and σ_{zz} respectively, where the common part of $-E/1-v^2$ are departed and the value at original point (0,0) of stress components σ_{rr}, $\sigma_{\theta\theta}$, σ_{zz} are infinite which are not be marked in plots.

4.2 Comparison of penetration depth, normal surface stress and applied force

For the penetration depth indented by conical and spherical profile of micro/nano-robot, by substituting the profile of $z=r$; $z=0.5r^2$ into the solution of dual integral and considering the boundary condition $\sigma_{zz}(r=a, z=0)=0$ for this type of incomplete contact, as the obtain of Equation (14) then the following relationships can be obtained,

$$D^{Conical}=\frac{\pi\cdot a\cdot k}{2},\quad D^{Spherical}=\frac{a^2}{R} \tag{26,27}$$

Furthermore, by substituting the Equations (24,25) into the normal stress components and setting $z=0$ in Equation (12), the normal surface stress distribution indented by conical and spherical profile can be computed respectively as below,

$$\sigma_{zz}^{Conical}(r,0)=\frac{-E}{2(1-v^2)}\cdot k\cdot \mathrm{Cosh}^{-1}(\frac{2D}{\pi\cdot k\cdot r}),\quad \sigma_{zz}^{Spherical}(r,0)=\frac{-E}{(1-v^2)}\cdot\frac{2}{\pi\cdot R}\cdot(D\cdot R-r^2)^{1/2} \tag{28,29}$$

Having the stress distribution of normal surface as above, the applied force can be easily deduced that based on the Equation (22), thus,

$$P^{Conical} = \frac{E}{\left(1-v^2\right)} \cdot \frac{2}{\pi \cdot k} \cdot D^2, \quad P^{Spherical} = \frac{E}{\left(1-v^2\right)} \cdot \frac{4}{3} \cdot R^{1/2} \cdot D^{3/2} \qquad (30,31)$$

Immediately, those results of penetration depth D, normal surface stress distribution and applied force P also can be found are identical with using the known formulas[14]. Figure 3 illustrate the numerical results of above relationships where the same contact radius a=1, and penetrating depth D=1 is fixed respectively.

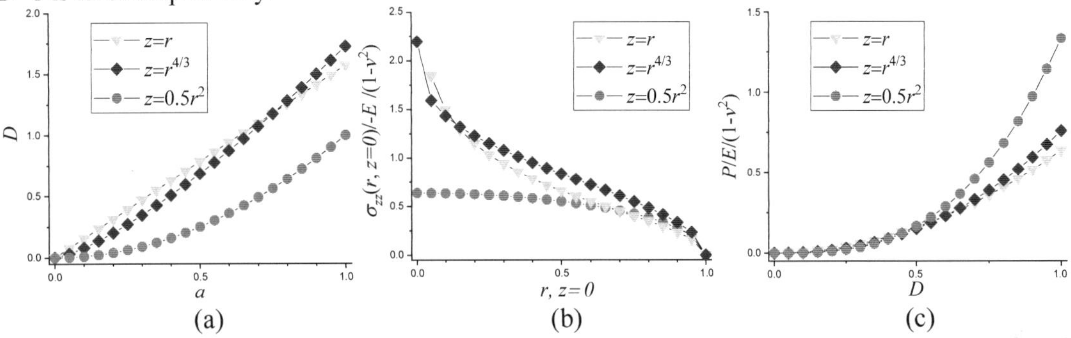

Figure 3. Elastic responses of cell indented by three different profile of maicro/nano-robot, where the generatrix of axisymmetric indenter is $z= r$; $z=r^{4/3}$; $z=0.5r^2$ respectively. (a) show the relationship between penetration depth D and contact radius a; (b) indicates the normal stress distributions on the surface of membrane which proving a quantitative indicator to understand the critical rupture stress of cell membrane, and the common part of $(E/1\text{-}v^2)$ are departed from plot (b) and (c); (c) presents the relationship between applied force P and penetration depth D which can be used in related field like measurement of modulus of softmatter, and P also corresponding the required force output from micro/nano-robot.

5. Conclusions

In this paper, the minimal resistance profile of micro/nano-robot firstly be derived when it locomoting in the fluid environment of vascular systems. Originating from the Bernouli's equation, combining the variational method with the one-dimensional Euler-Lagrange equation, the profile of 4/3 order curve of generatrix of axisymmetric body was obtained. Secondly, the elastic responses of the cell are studied as it was penetrated by the micro/nano-robot, which based on the Hankel transform and the elementary solution of Sneddon for solving contact mechanics. The normal surface stress distribution on the cytomembrane and the applied load P are derived respectively, which the former variable $\sigma_{zz}(r, z=0)$ indicates the critical rupture stress of cytomembrane and the latter one P corresponding the required force that the micro/nano-robot shall output to penetrate the cell membrane. Especially the interior stress and displacements distribution of the cell are explicitly derived for the first time as we known.

Acknowledgments

The authors would like to acknowledge the partial supports provided by the National Natural Science Foundation of China (No. 51575090, No. 11272083, No. 11502049), the Fundamental Research Funds for the Central Universities (No. ZYGX2014Z004, No. ZYGX2016KYQD118, No. ZYGX2015J084), the China Postdoctoral Science Foundation Grant (No. 2016M590873) and the National Youth Top-Notch Talent Support Program.

References

[1] Senju S, Koba C, Haruta M, Matsunaga Y, Matsumura K, Haga E, Sasaki Y, Ikeda T, Takamatsu K and Nishimura Y 2014. Application of iPS cell-derived macrophages to

IOP Conf. Series: Journal of Physics: Conf. Series **1209** (2019) 012019 doi:10.1088/1742-6596/1209/1/012019

cancer therapy. *Oncoimmunology* **3** 1.

[2] Yu J, Hu K, Smuga-Otto K, Tian S, Stewart R, Slukvin I I and Thomson J A 2009. Human induced pluripotent stem cells free of vector and transgene sequences. *Science* **324** 797–801.

[3] Bian S T, Zhou Y C, Hu Y W, Cheng J, Chen I, Xu Y C and Liu P 2017. High-throughput in situ cell electroporation microsystem for parallel delivery ofsingle guide RNAs into mammalian cells. *Sci. Rep.* **7** 42512.

[4] Kang W, Yavari F, Minaryjolandan M, Giraldovela J P, Safi A, Mcnaughton R, Parpoil V and Espinosa H D 2013. Nanofountain Probe Electroporation of Single Cells. *Nano Lett.* **13** 2448.

[5] Wen P. J. et al 2016. Actin dynamics provides membrane tension to merge fusing vesicles into the plasma membrane: *Nat. Commun.* **7** 12604.

[6] Bahrami A H, Lipowsky R and Weikl T R 2016. The role of membrane curvature for the wrapping of nanoparticles. *Soft Matter* **12** 581.

[7] Lin X, Wu Z, Wu Y, Xuan M and He Q 2016. Self-Propelled Micro-/Nanomotors Based on Controlled Assembled Architectures. *Adv. Mater.* **28** 1060–1072.

[8] Gao W, Dong R F, Thamphiwatana S, Li J X, Gao W W, Zhang L F and Wang J 2015. Artificial Micromotors in the Mouses Stomach: A Step toward in Vivo Use of Synthetic Motors. *Acs Nano* **9** 117–123.

[9] Dai B H, Wang J Z, Xiong Z, Zhan X J, Dai W, Li C C, Feng S P and Tang J Y 2016. Programmable artificial phototactic microswimmer. *Nat. Nanotechnol.* **11** 1087.

[10] Singh A V and Sitti M 2016. Targeted drug delivery and imaging using mobile milli/microrobots: A promising future towards theranostic pharmaceutical design. *Curr. Pharm. Des.* **22** 378–388.

[11] Elger D F and Roberson J A 2013. *Engineering fluid mechanics*. Wiley Hoboken (NJ).

[12] Courant R and John F 2012. *Introduction to calculus and analysis I*. Springer Science & Business Media.

[13] Sneddon I N 1995. *Fourier transforms*. Courier Corporation.

[14] Sackfield A, Hills D A and Nowell D 2013. *Mechanics of elastic contacts*. Elsevier.

AUTHOR INDEX